A HISTORY OF THE MURRAY CANAL

Murray Canal looking west

Murray Canal looking east

A HISTORY OF THE MURRAY CANAL

DAN BUCHANAN

One Printers Way
Altona, MB R0G 0B0
Canada

www.friesenpress.com

Copyright © 2024 by Dan Buchanan
First Edition — 2024

The image on the front cover was taken by the author.

The two aerial images of the Murray Canal were produced by Sean Scally and appear with his permission. https://www.youtube.com/channel/UCNYUFiePG4r2pkpcyw7pfrA/videos

All rights reserved.

No part of this publication may be reproduced in any form, or by any means, electronic or mechanical, including photocopying, recording, or any information browsing, storage, or retrieval system, without permission in writing from FriesenPress.

ISBN
978-1-03-919526-4 (Hardcover)
978-1-03-919525-7 (Paperback)
978-1-03-919527-1 (eBook)

1. HISTORY, CANADA, PROVINCIAL, TERRITORIAL & LOCAL, ONTARIO (ON)

Distributed to the trade by The Ingram Book Company

TABLE OF CONTENTS

Dedication ix
Preface xi
Acknowledgements xiii
The Powles Report xiv
Chapter 1 Simcoe's Canal 1
Chapter 2 Early Surveys and Advocates 5
Chapter 3 Optimism and Lobbying 8
Chapter 4 In the Halls of Parliament 12
Chapter 5 The Final Push 18
Chapter 6 Contracts and Opening the Works. 23
Chapter 7 Building the Canal 28
Chapter 8 Demonstration and More Work 35
Chapter 9 "Canalis" Tours the Works in 1888 38
Chapter 10 Final Work and Opening. 42
Chapter 11 Operations and Improvements 47
Chapter 12 Fixing and Changing 51
Chapter 13 Problems with Bridges 56
Chapter 14 People of the Canal. 61
Chapter 15 The Murray Canal Today 70
Notes 77
Appendix A: Order in Council Approving the Route 95
Appendix B: "Canalis" Tours the Works in 1888 97
Appendix C: Has Shovel Used To Open Work On Murray Canal 100
Appendix D: A Bit of First-Hand Information with Local Geography 102
Appendix E: The Naming of The Carrying Place Post Office. 103
Appendix F: Land Expropriation Records 105
Appendix G: Supporting Material 108
Appendix H: Note from Colin Powles 109
Bibliography 110
Illustration Credits. 112
Index 116

DEDICATION

On April 8, 2023, a big party was held in Brighton celebrating the one hundredth birthday of Florence Chatten. Friends and family greeted the birthday girl, who was delighted to see so many old acquaintances. Music was the order of the day, as Florence's family and musical friends had arranged for an afternoon of tunes and fun.

I, the author of this book, am lucky to count Florence Chatten as a good friend. My experiences with Florence began well before I came back to live in Brighton in 2010, but escalated after that as history activities increased in the community. Florence is always positive and encouraging to everyone, and her love of local history meant that she acted as a mentor for The History Guy. Her history book about Brighton Township added an important expression of local history to the community and her participation in many events at Hilton Hall Heritage Centre and at the larger History Open House events helped bolster a wider knowledge of local history.

I value very highly the hours spent talking with Florence about the history stories of the area. Her insight and knowledge have been instructive and her humorous nature always much appreciated. I look forward to many more visits with Florence so we can talk more history.

Florence Chatten 2018

PREFACE

A curious phenomenon developed over several decades as I became involved in researching and telling history stories. It seemed odd that there was no history of the Murray Canal. Sure, it was mentioned here and there, but I could not find any comprehensive document that explained how and why the decision to build the canal was finally reached and how the construction happened.

That is until I was handed a document that contained exactly that. Well, almost. The Powles Report, which is described in more detail on a following page, showed me all the detail I ever wanted to see about the long process that led up to the building of the canal as well as plenty of detail on its construction and operations. For a time, I sort of wallowed in this intense historical detail, indulging my instincts as a historical animal.

However, I also began to feel that this information should be presented to the public. It was a story worth telling. My experience with writing and publishing three history books and doing lots of history projects gave me the tools and the confidence to compile a history of the Murray Canal. It seemed apparent that, with the underlying detail from the Powles Report, a history of the Murray Canal was a feasible project.

Initially, I was focused more on the history of the Carrying Place, which is intimately related to the Murray Canal. In order to understand the development of the Murray Canal, you need to understand the function and evolution of the Carrying Place. This approach resulted in two separate presentations that I did at the Brighton Public Library: the first in October 2022, where I told the story of the Carrying Place; and the second in February 2023, where I told the story of the Murray Canal. In fact, this event was over-subscribed, so we had another one a month later and the room was full again.

Of course, a forty-five-minute presentation represents only a small fraction of the actual historical information I have collected on any given topic. It was not long before this collection was burning a hole in my virtual pocket. I began to feel that a more complete form of the story was required. It was also clear that the story should emphasize the Murray Canal with the Carrying Place as part of the early context. In spite of the complexities, difficulties, and expenses involved in publishing a book, it seems that I just can't help myself. When there is a very good story to tell, it's my job, so just do it.

The story that results may seem, in places, to be merely a regurgitation of the Powles Report.

Dan Buchanan, The History Guy of Brighton, Ontario

Three history books by Dan Buchanan

Presentations at Brighton Library

However, if you actually read that document, you will see it is a comprehensive historical report that is meant for a department head and not for public consumption. Much of the information is exactly what I wanted to use, but I had to make a story out of it, something that people could and would read. In many places, I include text in the form of quotes and, in some cases, the original text is presented in my own words.

However, this story is much more than that. My approach to telling a history story is always to provide lots of context, which lends meaning to events. The people in the story, in particular, should be familiar to us in a way that makes clear their motivations and explains their actions. In this case, the practical details from the Powles Report provide a great foundation and I have furnished the house, so to speak.

It may also be interesting to note that the copy of the Powles Report I received had no images at all, no maps or pictures or portraits, just text. My normal way of telling a history story in a forty-five-minute PowerPoint presentation took over from there and led me to add all the images I felt were required to support this story. I love maps, and there are many map segments, which are often enhanced with arrows and labels to highlight the message. Readers who don't like maps can ignore them, and those who do can use them as much as they like.

The larger format of this book is very much a departure for me. However, it more easily accommodates all the maps and pictures, many in color, which lend to the presentation of the historical story. I'm extremely proud of this book, and I hope readers will enjoy it.

No one book is ever the complete telling of a history, especially for something as complex as a canal. However, this story is offered to the community as an addition to the existing knowledge. Now we can say there is a history of the Murray Canal.

Dan Buchanan

Dan Buchanan
The History Guy of Brighton

ACKNOWLEDGEMENTS

While the information in the Powles Report provided a clear path forward for this book, the impetus for approaching the topic came from the War Claims Documents from the period after the War of 1812, which were brought to my attention by David Harris of Belleville. This information gave me a lot more knowledge about the Carrying Place and its people, including the additional documents supporting a brand-new version of the story of George Gibson's schooner, which can be seen on my web site.

Around the same time, Sean Scally, the talented and prolific producer of historical videos, began to speak to me about doing a video about the Carrying Place and the Murray Canal. We did the recording in 2019, and the video was released in April 2023. It is available on YouTube and a link is also on my web site.

With all of this, I thought I would try to gauge public interest in the topic by doing my normal kind of presentation. Thanks very much to the Brighton Public Library for providing a venue for three events, as mentioned in the preface. The interest was acute, so I was convinced that a book was necessary, even though a major undertaking.

Sean Scally also supported the work by providing two beautiful aerial photos of the Murray Canal as well as one of Portage Road in Carrying Place. Several other images came from Sean's collections as well.

Richard Hughes, of the Hastings County Historical Society, was helpful in locating documents and Amanda at the Community Archives of Belleville and Hastings County was quick to find documents.

Thanks to Christopher Kovach of Parks Canada for investigating the status of the Powles Report. In the end, he was able to provide permission for me to use the document to support this book.

The collections of the Library and Archives Canada were particularly useful in providing many of the Orders in Council that define the progress toward the building of the Murray Canal. Other documents, such as letters between politicians and parliamentarians, lend contemporary voices to the story.

The collections of the Brighton Public Library include documents in support of local history, and none more important than *The Tobey Book*, which has been digitized. The social and political issues of the times when the building of a canal was being discussed helped to frame the story in realistic terms.

XIV THE POWLES REPORT

Much of the detailed information about the Murray Canal that was available to the author came from Colin Powles's "A Construction, Operations and Maintenance History of the Murray Canal," aka the Powles Report, which was created in the summer of 1991. This is a hundred-page typewritten document that Parks Canada calls an internal white paper. It has no images but lots of references to original documents.

```
        A CONSTRUCTION, OPERATIONS AND MAINTENANCE HISTORY
                        OF THE MURRAY CANAL



                            Colin Powles
                        Canadian Parks Service
                        Ontario Regional Office
                            Summer 1991
```

Colin Powles worked for the Canadian Parks Service, which we now call Parks Canada. The introduction provides a clear statement of the purpose as well as the content of the document.

It says, "The primary purpose of this report is to outline the convoluted circumstances behind the decision to construct the Murray Canal… In addition to this, the following discussion will present a history of the construction, operations and maintenance of the Murray Canal."

This document has been analyzed thoroughly and many details included in this book are from that source. There are many quotes directly from these pages as well as many places where I present ideas from the report in my own words. While there is a lot more to the history of the Murray Canal than can be seen in this document, this is certainly the place to start.

I need to thank my friend Phil Spencer, who lives south of the canal, for passing a copy of this document to me several years ago.

In addition, many thanks to Parks Canada for investigating the status of this document and providing me with permission to use it for the purposes of creating this book.

Readers who are interested in seeing a digital copy of the Powles Report can download a PDF file from my web site at http://danbuchananhistoryguy.com/the-murray-canal1.html.

See Appendix H for "Note from Colin Powles," which describes his involvement in the creation of the report, presented in his own words.

CHAPTER 1
Simcoe's Canal

The idea of a canal between the Bay of Quinte and Presqu'ile Bay was discussed in official circles during the 1790s when Upper Canada's first lieutenant governor, John Graves Simcoe, was struggling to build the foundation of what he and others hoped would be a loyal and prosperous British colony.

Simcoe had been appointed as lieutenant governor of Upper Canada on September 12, 1791,[1] but it was not until June 24, 1792,[2] that he and his wife Elizabeth arrived in Upper Canada. Elizabeth Simcoe's diary says that the official party enjoyed the luxury of sailing on the *Onondaga*, a schooner that took them from Cataraqui (later Kingston) to Newark,[3] which was at that time the capital of Upper Canada. That means they did not cross the Carrying Place between the Bay of Quinte and Wellers Bay on that first trip, although we can expect that Simcoe had studied maps of Lake Ontario enough to see the critical nature of the small isthmus with the obvious name.

Lieutenant Governor John Graves Simcoe

One only needed to listen to experienced sailors on Lake Ontario and look at a map to realize that the west end of the Bay of Quinte was a very strategic place. Everyone knew that the waters at the east end of Lake Ontario were extremely treacherous and should be avoided. The Bay of Quinte and Wellers Bay provided a safe alternative.

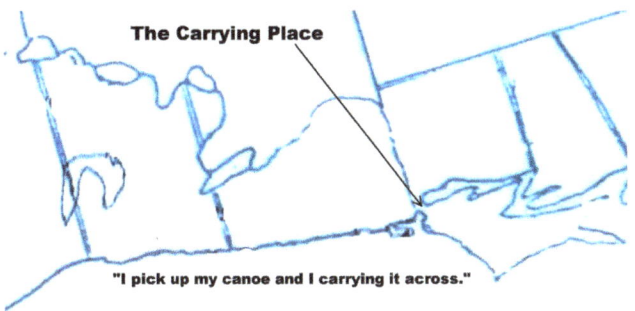

And the name was apt. Indigenous people from around the area had applied a practical name to this little isthmus because it was where you had to pick up your canoe and carry it across to the other side.

Nothing much had changed in this regard after Europeans came to the area. Everyone, whether servant or royalty, had to disembark at this point and carry the craft they sailed across to the other side before continuing along the north shore of Lake Ontario.

Much of the daily transportation of goods and people across the north shore of Lake Ontario employed canoes and batteaux, which hugged the shoreline in order to avoid rough water.

Flotilla of these small craft left Cataraqui and ran up Adolphus Reach, into the Bay of Quinte. Then, at the far western end of the bay, they were confronted with a small strip of land that everyone called the Carrying Place.

Simcoe was steeped in the art of military administration, especially in the engineering required to move armies and supply them properly. He grew up in England during the Golden Age of Canals, so he would be well aware of the benefit to be gained from the right canal in the right location.[4]

The idea of a canal took official form at a meeting in the council chamber at York on July 18, 1796, when it was "resolved, that three thousand Acres in the front of Murray be reserved for the purpose of facilitating the cutting of a Canal between the Bay of Quinte & Newcastle (or Presqu'isle) or for such other public benefit as it may be appropriated."[5]

This demonstrates that they were serious enough about a canal to set aside a large block of land to allow a canal to be built and establish a fund of land to pay for the work. It also shows that the name "Newcastle" was commonly used for the future town on Presqu'ile Bay, even in 1796.

Unfortunately, Simcoe left Upper Canada a few weeks later,[6] before any work on a canal could take place. However, for decades afterward, the idea of a canal at this location was kept alive with the expression "Simcoe's Canal."[7]

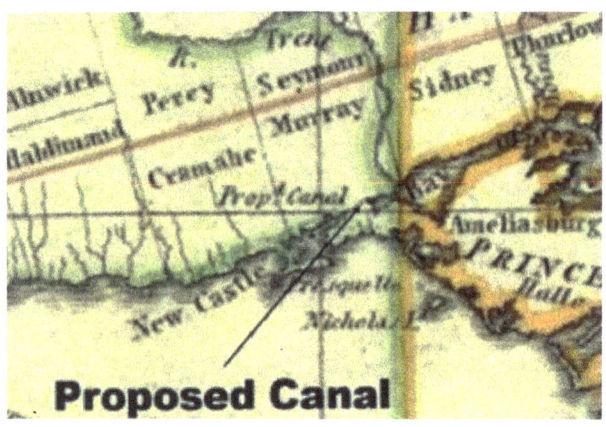

Map of Upper Canada, 1800

The later years of the 1790s saw intense surveying across the province of Upper Canada with the objective of providing organized townships where settlers could establish new homes. In order to document all of the new settled areas of the province, the General Surveyor's Office at York was charged with creating a map of the province showing every county, township, and town that existed at that time. The result was a beautiful and comprehensive map that shows us the progress of settlement as of the year 1800.[8]

Using modern technology, we can expand part of the map to show the details of the area of the counties of Prince Edward, Northumberland, and Hastings. The Bay of Quinte is clearly marked along with "Presque Ile" and the town of "New Castle."

However, looking closely at the south end of Cramahe and Murray townships, it is not difficult to see the words "Prop'd Canal" written on the map. Nearby we can see a thin black line drawn directly from the west end of the Bay of Quinte to the eastern point of Presqu'ile Bay.

Here is an official government document from 1800 that illustrates plainly that the idea of a canal from the Bay of Quinte to Presqu'ile Bay was already well established. In fact, it was very close to the exact route of the Murray Canal today.

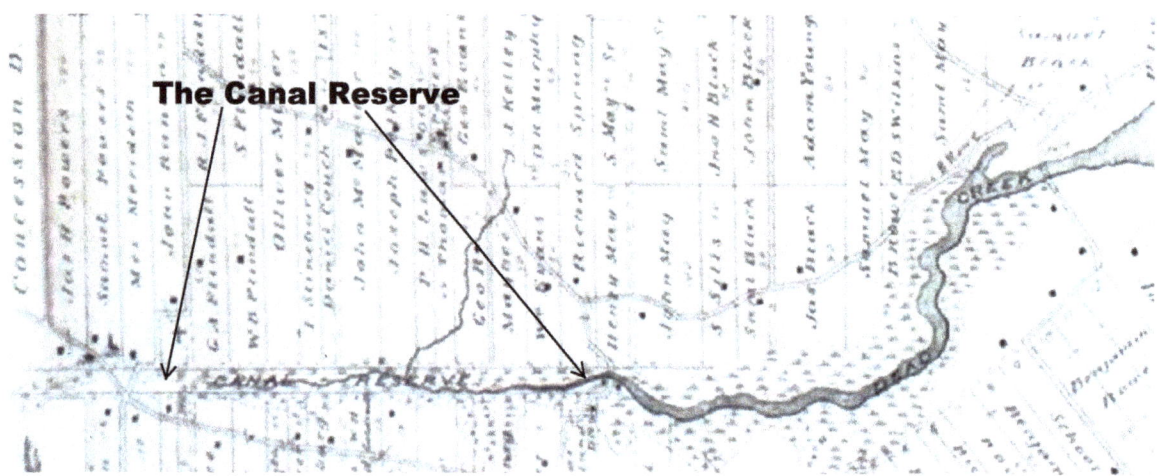

Canal Reserve, 1878 County Atlas Map

Tangible evidence of plans for a canal can be seen in the Belden County Atlas Map for Murray Township. These maps were created in 1878 and are a critical source of information for that specific time as well as for events and circumstances well back in time. Here is a snip of that map showing the mouth of Dead Creek on the right, coming west from the Bay of Quinte. The creek made a turn to the south, then flowed almost directly west. The western extent of Dead Creek is labeled "Canal Reserve."[9]

The Powles Report explains how this happened. "When the township was first surveyed in the late 18th century, 3,000 acres of land were set aside to pay for the cost of building the canal, with an additional 64-acre strip designated for the route of the cut itself."[10] The "Canal Reserve" shown on this map is that sixty-four-acre strip of land where the canal was expected to be built.

Among the first settlers at the Carrying Place was Asa Weller, who built a trading post at the west end of the portage path, on the north side, in Murray Township. Very quickly, he identified a business opportunity aside from the trading post. He established a batteaux portage service, which utilized teams of oxen to drag the flat-bottomed batteaux across the sand of the Carrying Place, charging $4.00 for each crossing. It was a simple but effective service.[11] Customers could complain about the fee, but what was your alternative? Needless to say, Asa Weller became prosperous because of the increasing traffic that needed his portage service.

The War of 1812 placed a magnifying glass on the Carrying Place, as regiments of soldiers and tons of supplies had to be passed along the north shore of Lake Ontario to support the armies to the west. The military brass complained long and hard about the serious delays in their operations and the excessive cost incurred. They had to empty each batteau, carry the contents across the portage, and then drag the batteau itself across to the other side.

Asa Weller's batteau portage service

Asa Weller did very well with his batteaux portage service. With the massive increase in traffic, he had to acquire more oxen and hire farm boys from the area to handle the volume. This would have been the right time to have a canal.

The experience of those living and working at the Carrying Place during the War of 1812 might be explained to some degree by reviewing a loss claim that Asa Weller sent to the government in 1815.[12] The list of losses includes 2,500 fence rails, which were used as firewood by the troops; one hundred fruit trees, which were easy picking for the soldiers; four tons of hay; and a barn turned into a barracks and then burned. The war had been profitable for Asa Weller, but also disruptive and destructive.

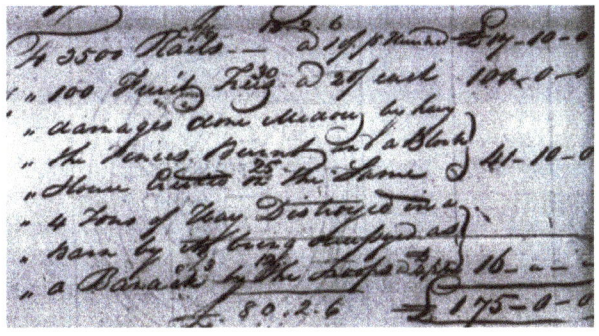

Asa Weller's War of 1812 Loss Claim

After the war was over, life along the Carrying Place returned to normal. Asa Weller's portage service would have competition from the upgraded York Road, which now went directly east from Brighton and did not follow the route of the old Danforth Road down into the county. The traditional function of the Carrying Place would gradually fade away for the regular traveling public.

Colonel Richard Bullock

Nothing had been done regarding the building of a canal, and by 1815, most of the land that had been set aside to pay for a canal had been distributed to settlers, leaving only the original sixty-four acres set aside for the actual canal.[13] The demand for land was too high, and the temptation to allocate the land for Crown grants or sell for cash was just too great. The rather nebulous idea of Simcoe's Canal was very quickly submerged in the rush to settlement.

Some of the land in this area would be granted to British soldiers who decided to stay in Canada after the war of 1812 was over. Colonel Richard Bullock was the most prominent of these, acquiring all 200 acres of Lot 15, Concession B, Murray, Township,[14] the location where the Danforth Road crossed Dead Creek on the way south to the Carrying Place. A significant number of fellow soldiers would settle in the area north of the Canal Reserve and the original east-west trail through that area would be called the English Settlement Road.[15]

Timber raft

After the war, a completely different group off to the west in Newcastle District still used the Carrying Place because they had no good alternative. The timber trade was very lucrative for the next several decades until sawmills across the landscape made lumber much more feasible. But, for now, large pieces of timber were lashed together to form massive rafts, which could be floated along the shoreline with the objective of reaching Montreal, where the timber could be sold for a good return.

Painful experiences had taught this group that they should not float their rafts around the south side of Prince Edward County since the unpredictable waters at the east end of Lake Ontario could destroy their rafts and endanger the lives of everyone on board. Their only alternative was to float the rafts into Wellers Bay, up to the west end of the Carrying Place. At this point, they completely disassemble the rafts and pulled each piece of timber across the portage. On the Bay of Quinte, the raft would be reassembled for the trip on to Montreal. This process was costly and time-consuming, but better than risking the angry waters to the south of Prince Edward County.[16]

Henry Ruttan of Cobourg

A solution to this problem was very evident to Henry Ruttan of Cobourg, who was a member of the Assembly for Northumberland County. In 1823, he was the first one to move for a survey of the lands at the Carrying Place with the objective of determining the best route for a canal.[17] An Act of Parliament authorized Mr. Macauly to conduct the survey, which concluded that a canal between the Bay of Quinte and Presqu'ile Bay could be built at a reasonable cost. Unfortunately, no action was taken.[18]

Here was an early example of the process that became very familiar. There would be lobbying, a survey, and then no work done to build a canal.

CHAPTER 2
Early Surveys and Advocates

Across Upper and Lower Canada, major canal projects were undertaken in the 1820s. The Lachine Canal west of Montreal opened in 1825.[1] The Rideau Canal from Bytown (Ottawa) to Kingston was opened in 1832.[2] The first Welland Canal opened in 1829, and by 1833, major expansion was already underway.[3]

These projects were funded by the government with two objectives in mind. First, they wanted to encourage more settlers to come to the Canadian colonies. Second, they wanted to provide efficient pathways for both the trade and defense of the country. Massive debt was incurred by Upper Canada in pursuit of these large and expensive projects.

Colonel By at Rideau Canal construction

John Colborne became lieutenant governor of Upper Canada in 1828[4] and would have a profound impact on the development of the colony. He was moderate and careful in his dealings with politicians, from the powerful members of the Family Compact to the loudest voices of the reformers. His approach was to concentrate on practical matters such as roads and bridges so people could see tangible improvements from their government. He had a folksy manner that allowed him to communicate with the farmer in the field, even though he was the king's representative. This was a significant departure from the arrogant and derisive stuffed shirts that had preceded him in the role.[5]

Sir John Colborne

An article in the *Kingston Chronicle and Gazette* dated November 1, 1834, expounds on the benefits of a canal between the Bay of Quinte and Lake Ontario. It explains the steps that had been taken up to that

Kingston Chronicle & Gazette, November 1, 1834

point toward that end and shows that the habit of surveys and no canal was well understood in 1834.

The article also sings the praises of Lieutenant Governor Sir John Colborne:

> Under his wise and energetic Administration, we have observed plans proposed for connecting the great waters of Upper Canada and for opening up new channels to agricultural and commercial enterprise, which are all upon a scale of such grandeur and extent, as literally to fill the mind with the most cheering anticipations, with respect to the speedy advancement of this magnificent Province in Wealth and importance. Among these plans, one for connecting the waters of the Bay of Quinte with those of Lake Ontario has been

revived, and the result was that another survey, authorised by Act of Parliament, took place last autumn.[6]

Nicol Hugh Baird

The survey that took place in 1833 was undertaken by Nicol Hugh Baird, a very active civil engineer. He had worked on the Rideau Canal under Colonel By and was also known as the designer of the covered bridge over the Trent River at Trent Port (Trenton) and the Presqu'ile Point Lighthouse.[7]

Baird recommended a canal between Twelve O'Clock Point on the Bay of Quinte and the north-east shore of Wellers Bay. He also included a lock in the canal as a means of coping with the fluctuation in water levels between Wellers Bay and the Bay of Quinte. In spite of Baird's work, no further action was taken.[8]

Rebels marching down Yonge Street

The next obstacle to development was the Mackenzie Rebellion of 1837 and 1838. The rebellion froze funding for all major projects and the colonies of Upper and Lower Canada experienced severe economic and social disruption.[9] At Cornwall, the Cornwall Canal was eighty percent complete, and the work just stopped. For several years, the unfinished canal was a big muddy ditch as the community around it slid into recession.[10]

After the rebellion was over, the British government sent Lord Durham to the colonies with instructions to talk with the people of the Canadian colonies. The politicians in London wanted to find out what the colonists were angry about so changes could be made that might prevent another rebellion. The result was one of Canada's most famous documents, the Durham Report.[11]

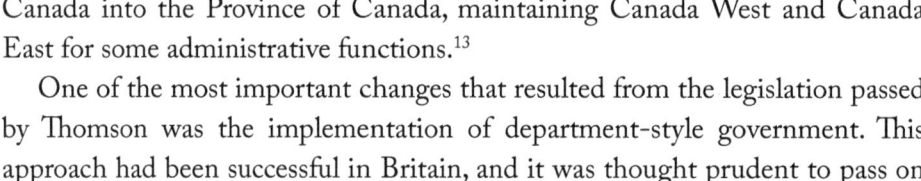
Lord Durham

Lord Durham's Report recommended sweeping reforms to the governments of Upper and Lower Canada. The obvious need was to modernize the structure of government, but there was also the underlying desire to move away from the traditional tight control of government by the elites of the colonies. The Family Compact would be forced to relinquish control to a more inclusive society, although full responsible government was not on the table just yet.

Charles Poulett Thomson

Charles Poulett Thomson was appointed as the next governor general for the British North American Colonies in the fall of 1839 and his primary task was to pass legislation in the parliaments of Upper and Lower Canada that reflected the recommendations in the Durham Report.[12] The new system took effect in February of 1841, highlighted by the major step of merging Upper and Lower Canada into the Province of Canada, maintaining Canada West and Canada East for some administrative functions.[13]

One of the most important changes that resulted from the legislation passed by Thomson was the implementation of department-style government. This approach had been successful in Britain, and it was thought prudent to pass on proven methods to the colonies.

One of the most important departments that was created at this time was the Board of Works, which would be responsible for what we call public infrastructure, including roads, bridges, canals, and, within a decade, railways.[14]

This new department was made up of professionals experienced with large infrastructure projects and hiring would be based on merit. Even more important, funding for infrastructure projects would be dedicated by the government, through the Board of Works, in order to ensure a good quality product and timely completion. This was a huge departure from earlier methods, and it was badly needed.

During his visit to the colonies, Lord Durham commissioned Lieutenant Colonel E. Phillpotts, royal engineer, to undertake a survey of the St. Lawrence River and the existing canal systems so that "the Commerce of the Canadas may be incalculably extended, and their general interests advanced to the highest pitch of prosperity."[15]

Statute creating the Board of Trade

Phillpotts agreed with Baird's route to Wellers Bay, but he also felt that Presqu'ile Bay was one of the best harbors in the area and should be developed as a "harbor of refuge."[16] Lord Durham had a broad view of transportation policy and tended to use rather vague terms such as "general interest" to promote the idea. However, for most people in Upper Canada, the main reasons for a canal tended to emphasize the safety of marine travel as well as a military function in light of aggressive tendencies from our neighbors to the south.

This flurry of activity did not result in shovels in the ground, but there was a further survey conducted in 1845 by a Mr. Lyons of the Board of Works.[17] This survey countered Baird's Wellers Bay route by recommending the Presqu'ile Bay route. The Lyons survey also said that a lock in the canal was not necessary. More importantly, advanced methods of testing the geology of the area resulted in the discovery of significant deposits of rock just below the surface along the north shore of Wellers Bay. This would make Baird's route untenable because it would dramatically increase the cost of digging a canal.[18]

As usual, nothing happened after the Lyons survey, although the geological findings would be foundational in future surveys.

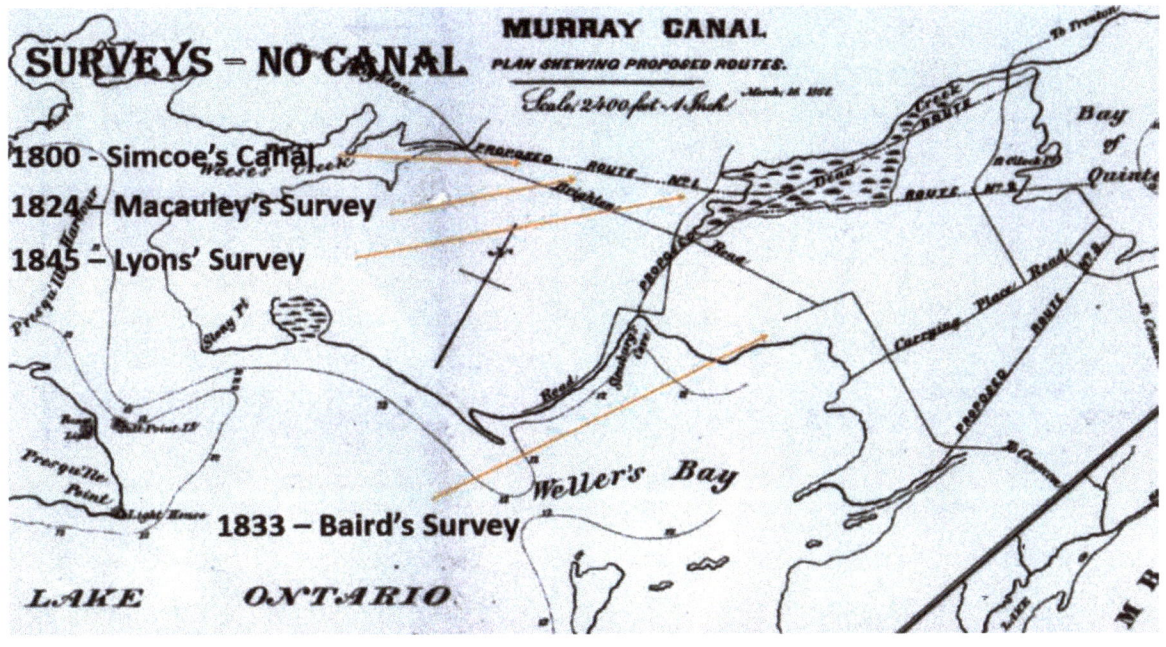

Lots of surveys but no canal; map showing routes surveyed up to 1845

CHAPTER 3
Optimism and Lobbying

On September 7, 1850, a very enthusiastic public meeting was held near Trenton, where many voices were raised in support of a canal. In that meeting, "it was resolved that the Reeve of Murray Township should contact officials from the surrounding townships with the purpose of petitioning the government to undertake the works. For the next several years petitions arrived periodically at the Legislative Assembly."[1]

The population of Upper Canada had doubled between 1830 and 1840. Settlement along the north shore of Lake Ontario had been completed to a large degree, and farmers raced to clear their land to grow wheat and other agricultural products that were in high demand. In the early 1850s, the export of lumber to New York State was at its peak. Government played its role by reducing many of the stifling tariffs that had been designed to protect the growing colonies. Lumber merchants along the north shore of Lake Ontario had major markets opened up for them on the south side of the lake as a building boom heated up in New York State. Industrialization was creeping into society as farmers' sons gravitated to the growing towns to take up trades. Growth and prosperity meant much greater demand for all kinds of consumer goods.

Farmers saw a canal as a quicker and easier way to move produce to markets. Manufacturers wanted reliable transportation of raw materials to their factories and finished product to markets. It was starting to make a lot more sense to a lot more people that Simcoe's Canal was a good idea.

During the 1850s, folks around Brighton were not unique in the region around the Bay of Quinte in experiencing a major spurt of growth and optimism. Brighton Township was created on January 1, 1852, providing a much better situation with a central town and a rural region with easy access to commercial establishments, schools, and social organizations.[2]

Railway construction crew, c. 1860

Employment was sky high with two major construction projects underway at the same time. The Grand Trunk Railway was being built from Toronto to Montreal, and the tracks ran just south of the town of Brighton. This was a very significant development that would open the small towns along the lake and the farming regions of the hinterlands to the wider world.

At the same time, the main road north from Brighton was being built. The Old Percy Road had been obsolete long before the floods of the Breakaway damaged an important part of the road near Hilton in April 1852. Now that the new reform government had provided incorporation of all municipal entities, Brighton Township had the financial tools to do a major infrastructure project. The Brighton and Seymour gravel road would run from the north shore of Presqu'ile Bay, directly north from the town, up through the middle of the new township, to meet the road from Campbellford. This resulted in a much better north-south road through Brighton Township. Another major project was the construction of a long wharf at the south end of the road, which would provide schooners from the lake convenient access to lumber and other produce from the area.[3]

Like their counterparts in Belleville and Trenton, the people of Brighton wanted the Murray Canal built as soon as possible. Alex Begg, editor of the Brighton newspaper, minced no words when he said that the canal "…ought to have been undertaken and completed many years ago, and it [was] truly incomprehensible why it [had] been neglected."[4]

In 1852, the Gilmour Lumber Company built a sawmill on the Trent River, beginning a half century of employment and industrial activity in the area. We should not underestimate the strength of Gilmour's voice in the lobbying for a canal.[5]

Gilmour Lumber Company of Trenton

Around the same time, all of the communities in the area took advantage of recent reforms that provided incorporation for every town, village, and township, giving municipalities more independence in financial dealings. The 1850s was a time of major upgrading of facilities such as harbors and wharfs, all with the objective of boosting the local economy.

In August of 1850, Belleville City Council appointed a committee to look into the issue of a canal and their report went to the Ontario Public Works Department, which was the new name for the Board of Works.

The report said, in part, "The inhabitants of this section of the country look upon this project as one of vital importance to their prosperity."[6]

The *Brighton Sentinel* concurred: "There is no single work of equal weight, as regards the interests of the trade on Lake Ontario, and perhaps eventually the defense of Upper Canada, which could be engaged in at the present moment."[7]

However, during the middle of the 1850s, the attention of the province was primarily on railways. The Grand Trunk Railway was built between Montreal and Toronto in the years 1853 to 1856. New train stations were built at every town along the way and people began to see in practical terms how this transportation link would change their lives.

Locomotive and wood car, c. 1860

Merchants could order the latest fashions from London or New York, and they would be delivered within a few weeks on a much more reliable schedule than steamship traffic could provide. Farmers now had a much better way to send their produce to the growing markets, and their children could take the train to attend high school in pursuit of a better education. A visit to the retail district of the larger towns became much less of an ordeal, completed in much less time and with a much greater degree of comfort and convenience. The train quickly became an integral part of daily life.[8]

Supporters of the canal were frustrated at this distraction and struggled to keep their advocacy before the eyes of the politicians. In 1846, Samuel Keefer, chief engineer of the Board of Works, had stated in a report about the proposed Murray Canal that "I have always looked upon it as a project designed for the especial benefit, accommodation and safety of steamboats and vessels navigating the Lakes, as a means of strengthening the defences of the country by affording, in time of actual hostilities, a secure inland route for vessels and munitions of war. Taking a more extended view, it would appear to be intimately connected with Welland-St. Lawrence Canals."9

A decade later, Mr. Keefer was commissioned by the Hastings County Council to report on the potential for a canal and his report raised an issue that was new to the discussion. He felt that the presence of various sand bars in the Bay of Quinte "would not permit navigation beyond a six foot depth."10 This was not what the advocates wanted to hear, especially those who felt that a canal could serve a military role.

Another point of contention in this period was the competition between Presqu'ile Bay and Wellers Bay in the role as harbor of refuge. The government formalized this term in order to provide support for sailors on Lake Ontario in bad weather. For many years, both of these bodies of water had been used as places of refuge in storms and the battle raged on the benefits of one over the other.

Presqu'ile Bay and Wellers Bay — harbors of refuge

Presqu'ile Bay was better protected from westerly gales on the lake, but its entrance was complex and difficult because of the shallowness of the Middle Ground Shoal and the difficulty of finding the old channel in rough weather.

Wellers Bay had an easier entrance but did not provide as much shelter from strong westerly storms. Also, there were growing examples of sand accumulating at the entrance during storms. This often led to dredging just to keep the channel open for navigation.

The debate continued on this question in the government as well as in taverns and on wharfs around Lake Ontario. Which bay you favored might depend on specific experiences you had in certain storms with either of the two bays. People who lived around each bay lobbied hard to have their home waters favored with the label "harbor of refuge."11

The 1860s brought the American Civil War and poor relations between Britain and the USA. In 1862, a commission was sent by the British government to review military facilities in the colonies. The commission's report suggested that the Bay of Quinte might be a good place for a significant naval and military facility. Of course, it would be necessary to have a direct water link to Lake Ontario, so a canal at the west end of the Bay of Quinte was a given for the military.12

Local residents around the Bay of Quinte were once more seized with the idea of a canal. In 1865, support for a canal came from a distant source in a report by the Montreal Board of Trade on Inland Navigation, which reported that the canal would be important to avoid the dangerous stretch of water of the Prince Edward peninsula, and that "the expense of the work would be small, while the advantage would be great."[13] Just what the local folks wanted to hear.

Toward the end of the Civil War, Canadians began to fear the prospect that the huge Union Army might simply turn itself north after it was finished defeating the South. It was a fear that motivated the politicians to gather together and present a strong front to the Americans, in the form of a federal union.

There were lots of other good reasons for a confederation of the British North American colonies, but national security was an effective motivator. Fear of future invasion from the south also led to the upgrading and enhancement of fortifications and the organization of militias.

Confederation would become a reality, but the report from the British military commission ended up on a dusty shelf. The fear of American aggression dissipated with the end of the Civil War. It became obvious that the Americans had no intention of turning north. Once the Fenian raids became a memory, any thought of the need for a canal for national defense purposes faded into history.

CHAPTER 4
In the Halls of Parliament

During the 1860s and the 1870s, the Murray Canal was a common topic of discussion in the halls of Parliament in Ottawa. Talk often revolved around two particular politicians, James L. Biggar and Joseph Keeler III. Both were members of Parliament for the riding of Northumberland East, at different times. Since the canal would be in their riding, they were strong advocates for the building of the Murray Canal.

James L. Biggar, MP

James L. Biggar (1824–1879) grew up in a prosperous merchant family on the Trenton Road at the west end of the Bay of Quinte. His father, Charles Biggar, had been born in Steuben County, New York, in 1797 and moved with his family in 1803 to Queenston Heights in the Niagara District of Upper Canada. His sister, Mary Biggar, had married Robert Hamilton, a son of the famous Robert Hamilton, trader and merchant at Queenston Heights in the very early years of the colony.

By 1816, 19-year-old Charles had established himself as a merchant at the east end of the Carrying Place and would thrive in spite of changes in the area that reduced the traditional use of the narrow isthmus from a critical part of the regional transportation system to a local road in a quiet rural community.

His son, James, attended Victoria College in Cobourg, a credential that would serve him well throughout his life. The family business would continue as James Biggar operated a very popular general store on the Trenton Road. He also participated extensively in real estate transactions, as was common for people who had cash on hand. His name can be seen routinely in the land registry records for all the surrounding townships. For many years, he held the position of post master for the Murray Post Office, which local folks would call the village of Carrying Place.[1]

Parliament Buildings of Canada, Ottawa

It was natural for a young gentleman in these circumstances to become involved in politics. He was elected to the House of Assembly for the Province of Canada in 1861, representing Northumberland East. As a member of Parliament, James L. Biggar was very active in lobbying for the building of a canal.[2]

While many politicians were engaged in discussions that would lead to a confederation of the colonies, Biggar pushed hard to educate members of Parliament about the need for what was now commonly called the Murray Canal, since it would be located in Murray Township.

The work he did during his first years as a member of the House led to his appointment as chairman of a Select Committee that was charged with looking into the issue of building the Murray Canal. On July 24, 1866, he submitted the committee's report, which said, in part:

> That under these circumstances Your Committee would recommend that a Survey be made of the neck of land lying between Lake Ontario and the Bay of Quinte, and also of the Harbours of Presqu'isle and Weller's Bay, for the purpose of ascertaining the cost and feasibility of said Canal, and that the Survey should be commenced with the least possible delay.[3]

As a result, a survey was undertaken during the spring and summer of 1867 by engineer J. H. Rowan, who made his report to John Page, chief engineer of Public Works. This was the most comprehensive survey to date, addressing three different potential routes for the canal and comparing them on the merits and costs.

The traditional route up Dead Creek to Presqu'ile Bay was studied in much more detail than in previous surveys. This was, of course, the longest route and therefore the most expensive. Even then, the estimates grew larger with the discovery of nearly two-thirds of a mile of rock formations at a depth of nine feet near the west end of the route. The depth of the canal was expected to be 11 feet, so this rock would present added cost for dredging.[4]

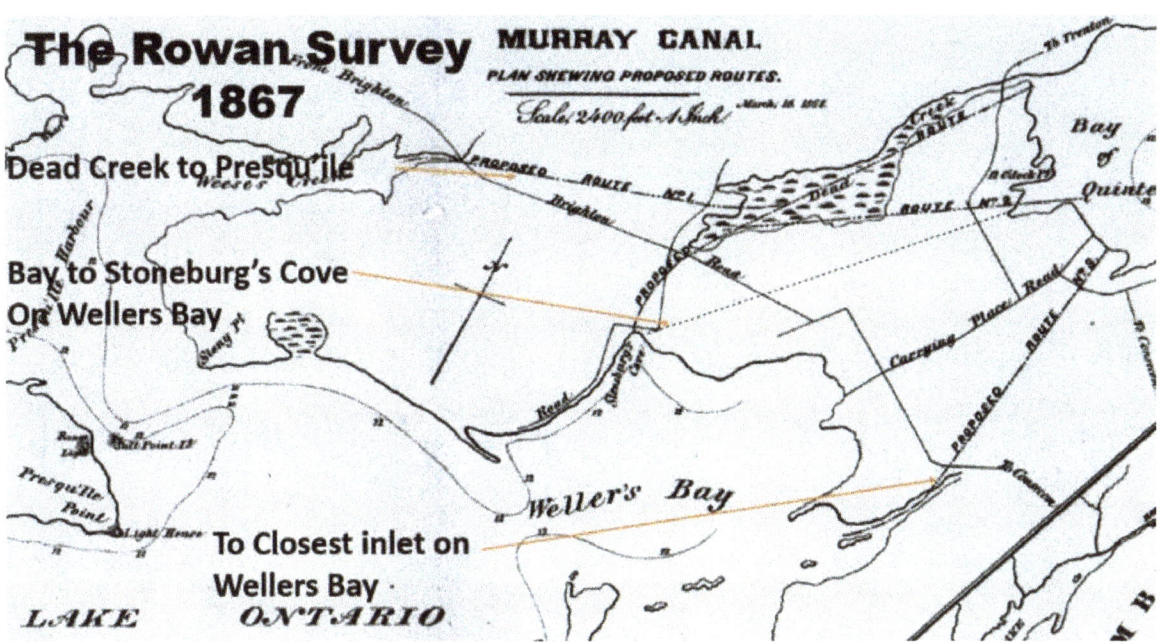

Rowan Survey 1867 studied three possible routes for the canal.

The second and much less expensive route was from Twelve O'Clock Point on the Bay of Quinte to Stoneburgh's Cove on the north shore of Wellers Bay. The third route was being considered for the first time, running from the Bay of Quinte a bit south of Portage Road, on a south-west angle to the nearest inlet on the east side of Wellers Bay. This was the only route contained inside Ameliasburgh Township, Prince Edward County, while the others were in Murray Township, Northumberland County.

Rowan was very clear in his report that the last two routes were more desirable, partly because they were considerably less expensive, but also because Wellers Bay already had access to Lake Ontario at the proper depth. If Presqu'ile Bay were the western terminus of the canal, extensive dredging of the large limestone shelf at the entrance to the bay would be required.

It may be useful to mention that this was before the government dredged a straight channel out through the Middle Ground Shoal off from the Presqu'ile Point Lighthouse, providing easy access to the bay from the lake. This was done to avoid the convoluted entrance into the bay, which had always been a dangerous and difficult factor in selling Presqu'ile Bay as a harbor of refuge. With the new channel, completed in 1872, the bay would receive higher status, which led to better funding for items such as range lights and further dredging.

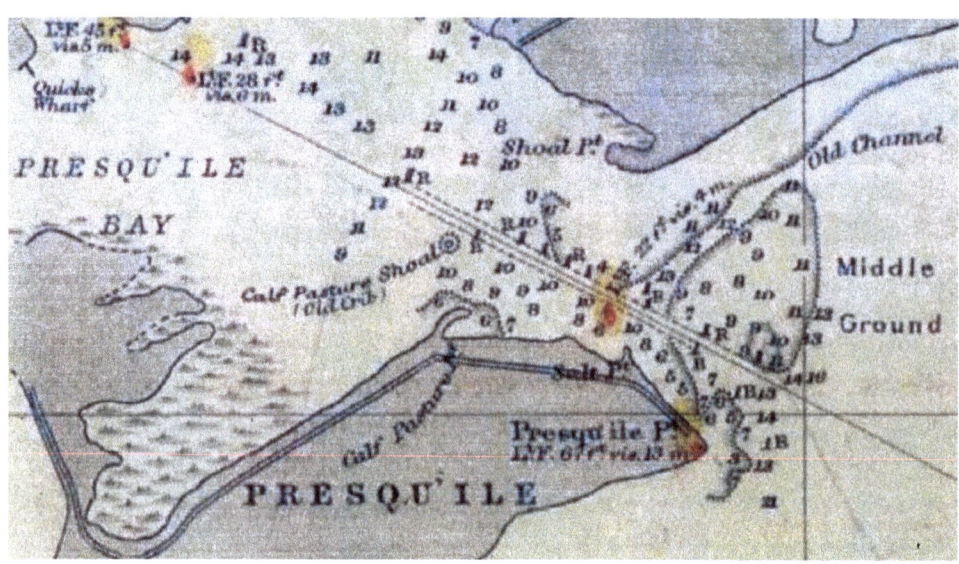

Presqu'ile Bay — channel through Middle Ground Shoal

Rowan also explained how the route through Presqu'ile Bay would be "tortuous," requiring ships to navigate several sharp turns in order to travel in either direction between the canal and Lake Ontario. His report also included the need for a lock to control water levels at each end of the canal. It was estimated that building this lock in a place where significant rock was just below the surface would be very costly, even for the shorter route to Wellers Bay.[5]

The language is interesting. Rowan seems to be rather non-committal about the route and appears to be more intent on presenting reasons for not building a canal. John Page, the chief engineer, was even more blunt when he said, "The reasons referred to as having been urged in support of this undertaking are entirely of a commercial nature, and although evidently of considerable importance, it may be questioned whether the advantages which the work (if executed) would confer upon the general navigation would warrant so large an expenditure."[6]

Sir John A. Macdonald

The Powles Report provides a theory to explain the reluctance of federal politicians, other than the main advocates, to support the building of the Murray Canal. Strong support around the Bay of Quinte region may have been countered by the fear in the minds of the members of the Kingston Board of Trade that the canal might reduce Kingston's advantage as the focus of transportation on the St. Lawrence River.

Newspaper articles discussed this idea at length and came to the conclusion that the member for Kingston, John A. Macdonald, was exerting influence in favor of his constituents by tamping down enthusiasm for the Murray Canal in the halls of Parliament. At least, if you asked a politically savvy resident on the Bay of Quinte, you would hear this theory.[7]

In the midst of all this, Confederation took place on July 1, 1867, with the Province of Canada reverting back to two separate provinces: Ontario and Quebec. These two provinces were joined with

New Brunswick and Nova Scotia to make up the original Dominion of Canada.

John A. Macdonald was the prime minister of the new dominion, and he set about strengthening the position of the new country in every possible way. His primary concern was to combat domination by American interests, and the best way to do that was to present American expansionists with a strong and determined neighbor to the north that could rely on solid backing from the most powerful military and political force on earth, the British Empire.

Confederation also brought Joseph Keeler III to the halls of Parliament. He was from Colborne and had won the seat for Northumberland East that James L. Biggar had occupied for several years. Everyone around Colborne knew that this fellow was "Little Joe," his father was "Young Joe," and his grandfather was "Old Joe."[8]

Joseph "Old Joe" Keeler (1763–1839) had been born in Berkshire County, Massachusetts, but his family were loyalists who moved to Vermont to escape persecution during the War of Independence. Early in the 1790s, Old Joe led a group of families from Vermont to the north shore of Lake Ontario, where they established a settlement near the current village of Lakeport.

After the construction of the Danforth Road in 1800, the next generation of Keeler men established a general store on the road just north of their settlement. This convenient four-corners location would evolve into a village, and during the 1830s, it would be named Colborne after John Colborne, the beloved lieutenant governor of Upper Canada.[9]

Joseph Keeler III

Joseph Keeler III was well positioned to play a key role in service to his community, and one way he did that was to promote the idea of a canal at the far corner of his riding in Murray Township. He saw a canal at this location as a benefit to all farmers in the region, including those as far west as his home in Colborne and on to Cobourg.[10]

Keeler was an avid supporter of the canal long before he gained a seat in Parliament as part of Macdonald's victory in the first election for the Dominion of Canada. Trade between Lakeport and Oswego, New York, had been lucrative for the Keeler family and others who sailed schooners across the lake full of lumber and other goods. The treacherous nature of this trip was embedded in the culture of the day, as everyone knew someone who had been lost in a storm on Lake Ontario.[11]

Once he was in the halls of Parliament, Joseph Keeler III began lobbying his fellow politicians on the issue, backed up by several petitions from local interests. These early efforts appeared to be making headway when the idea of a canal was address by the "Select Committee on Maritime and River Fisheries and Inland Navigation." The report from this body in 1869 was positive, saying, "It has the oldest and strongest claim of any work in the Dominion."[12]

However, the opinion of the "Royal Commission on Inland Navigation in Canada" in 1870 was much less positive. It reflected that in the 1870s, much like in the 1850s, canals were seen as secondary to railways, which should receive the greater share of government funding. Even then, one of the primary national interests was shipment of grain from the west, which had been cornered by the Erie Canal and various American railways. The Royal Commission emphasized the need to upgrade the existing Welland and St. Lawrence canal systems so that they could take a significant part of this trade.[13]

In this broad context, the Murray Canal was far down the list of priorities. The report said it was "entirely a work of local importance, and is not required by the general trade of the Dominion. In this view, while so many works of general importance are calling for execution, the Commissioners recommend that for the present the consideration of this canal be deferred."[14]

This would have been disappointing for supporters of the canal. In 1874, they would have much more to be concerned about. John A. Macdonald was caught in a scandal related to the funding of the trans-continental railway, so he and his government had to resign and an election was called. The Liberal Party came into power in this election, bringing James L. Biggar back to his position in the House as member for East Northumberland.[15]

Oddly enough, even though Biggar was a strong supporter for the canal, the politics of the day did not provide an opportunity for him to promote the cause. The period from 1874 to 1878 was very quiet as far as the Murray Canal was concerned.[16]

The federal election in the fall of 1878 changed the situation in several ways. The Conservative Party came back into power with a rush and Joseph Keeler was back in the House. He set to work immediately, re-establishing the issue of the Murray Canal in the halls of Parliament.[17]

While John A. Macdonald was once more the prime minister of Canada, he had suffered defeat in his old riding in Kingston. Eventually, he would win a seat in Victoria, British Columbia. The residents of the Bay of Quinte had been very conscious of the opposition to the canal from interests at Kingston, so this was seen as a very positive development.

The Powles Report provides the following from the *Daily Intelligencer*, admitting that "the limestone city would suffer [from the construction of the canal] to a considerable degree is not doubted; but we do not rejoice at such a prospect. Her influence has been potent hitherto to a certain extent in preventing the carrying out of this great work, but by her own act that influence has doubtless been materially weakened. We now have good ground for hoping that the canal will be made."[18]

James Lyons Biggar

This change in the political landscape encouraged supporters of the canal to make another concerted effort at lobbying. However, they would be without James L. Biggar, who had been the primary advocate for the canal in the halls of Parliament in the previous decade. He lost his seat to Joseph Keeler as part of the Conservative victory in 1878. Tragically, James Lyons Biggar passed away in New York State on May 24, 1879, leaving a family of eleven children, the youngest only seven years of age.[19]

The Bigger Picture

The Canals of Canada, by John P. Heisler

Canadian Historic Sites: Occasional Papers in Archaeology and History – No. 8

National Historic Sites Service
National and Historic Parks Branch
Department of Indian Affairs and Northern Development
Ottawa, 1973

The Canals of Canada

It may be useful to gain some perspective by looking at the content of a report that was done by John P. Heisler called "The Canals of Canada." This was an in-depth review of all the canals across Canada and their development over time. It was published in 1973 under the authority of the Department of Indian Affairs and Northern Development.

A search through this document finds very little mention of the Murray Canal. A map of "Trent Navigation and the Murray Canal" is found on page 33 and this on page 135:

> Yet another of the projects undertaken at this time to improve the St. Lawrence-Great Lakes navigation was the construction of the Murray Canal extending through the Isthmus of Murray and affording a connection westward between the headwaters of the Bay of Quinte and Lake Ontario, thereby enabling vessels to avoid the open lake navigation.[20]

Below this is a description of the building of the Murray Canal.

However, back on page 129, there is a paragraph that puts the Murray Canal in the wider perspective, although not mentioning it by name:

> The Canadian government accepted the royal commission's recommendations and decided to enlarge the canals; the scale of navigation in the St. Lawrence throughout was fixed at an available depth of 12 feet of water instead of 9 feet in the Lachine, Beauharnois, Cornwall, Farran's Point, Rapide Plat and Galops canals, and instead of 10-1/2 feet in the Welland. The dimensions of the locks on the enlarged canals were fixed at 270 feet between the gates and 45 feet in width instead of 200 by 45 feet. The least breadth of the canals at bottom was fixed at 100 feet. In 1873 this enlargement was authorized to be carried out on the Lachine and Welland canals and subsequently on the Cornwall Canal. Two years later, in 1875, strong representations were made to deepen the various channels for the passage of vessels drawing 14 feet of water.
>
> The government assented to this need for deeper channels and orders were given to place the foundations of all permanent structures on those parts of the works not then under contract at a depth corresponding to 14 feet of water on the mitre sills. These orders applied to all the principal works on the main line of navigation between Lake Erie and Montreal. Such dimensions would enable vessels of almost any ordinary build to pass carrying fully 1,000 tons burden.[21]

The second-to-last sentence contains the telling point. The Murray Canal was not considered a "principal work" in the broader navigational system. Further along in the document, on page 148, the point is made more clearly.

"Though the Murray, Trent, Rideau and Ottawa canals could be considered geographically as branches of the through east-west route, yet the operation of these canals served mainly a distinct local traffic."[22]

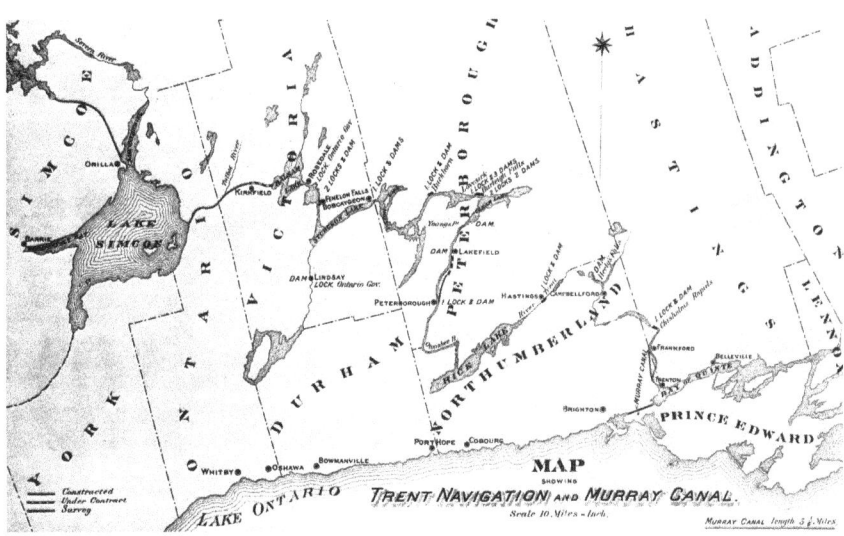

Trent Navigation and the Murray Canal

CHAPTER 5
The Final Push

In spite of the loss of J. L. Biggar, the fight for the canal took on new life. On October 20, 1879, a large meeting was held in Belleville by the Canal Committee of the Belleville City Council.[1] This meeting was attended by politicians, officials, and interested parties from all across the region. At this point, the emphasis was on creating a united front in support of the canal. Any petty local issues were to be submerged in a great united push to have the canal built.

At this meeting, the decision was made to establish direct communications between the Canal Committee and the government. Numerous petitions were set in motion and plans were made for a delegation of local politicians and members of Parliament to go to Ottawa and meet face-to-face with Charles Tupper, the minister of Railways and Canals.[2]

In hindsight, we can look back at these events and see that the meeting in Belleville in 1879 was the turning point in the campaign for the Murray Canal. They now had momentum.

In the middle of these developments, there was another marine disaster to add to the long list of reasons the Murray Canal should be built. On November 17, 1879, several dredges, derricks, and scows were being towed from Cape Vincent to Oswego when they were caught in a nor'easter.[3]

The result was a terrible disaster. Sensational newspaper reports in the early hours after the event reported upwards of thirty people lost. When the facts came to light in a few days, the final count was six dead, with several vessels lost. This event was used in political circles to drive home more urgently the point that the Murray Canal would reduce the loss of lives as well as ships and cargo in this dangerous area. Certainly, it caused a sensation in the House for a time and added to the already-building impetus to move ahead.

Marine disaster, November 17, 1879

On December 24, 1879, the delegation met with Charles Tupper in Ottawa. It was reported that Tupper expressed some surprise that the canal had not already been built. This indicates less his ignorance and more his political savvy in aligning himself favorably with the delegates at the outset.[4]

Then, in February of 1880, during the annual budget debates, a document called "A Paper on the Proposed Murray Canal" was presented to MPs and senators in Parliament.[5] It provided a brief history of the proposed canal and expounded on its potential benefits. Time had passed and the language of the advocates had changed to reflect current conditions. There was no mention at all of any military benefit, in spite of the fact that politicians would persistently use this talking point in their speeches.

The emphasis was now trade and prosperity. The lumber industry was a key supporter with record exports, much of it heading west. The issue of freight rates on the Grand Trunk Railway was included with the thought that the canal might force the railway to implement fair rates across the system.[6]

At this time, a new item was added to the list of potential benefits. Optimism was riding high for the iron ore industry in the north of Hastings County. The Blairton Mine at Marmora had been very productive from 1868 to 1873 and had closed in 1875, but operations were underway to open another and potentially larger mine at Coe Hill.

Up to that time, iron ore had been shipped by rail through the port of Cobourg. However, the movers and shakers around the city of Belleville dreamed that their harbor might be a main terminus for the iron ore industry and their plan required a canal to be built at the west end of the Bay of Quinte in order to ship product by water to the American smelters.[7]

The advocates for the canal put their feelings in raw political terms, speaking directly to John A. Macdonald's Conservative Party:

> Now, if there are two sections of the Province of Ontario that deserve especial consideration at the hands of the present Dominion Administration, they are the city of Belleville and the North Riding of the County of Hastings. The former has done much for Sir John and his friends, while the latter has for many years returned to Parliament a gentleman who now occupies a seat in the Cabinet — the Honourable Mackenzie Bowell. What have the city of Belleville and the North Riding of Hastings received in return for their loyalty to the Conservative cause?[8]

Joseph Keeler insisted that existing plans for work on the canal be presented to the House as a tactic to keep the canal top-of-mind for his fellow members. Mr. Bowell, who was in the cabinet, put his weight behind the project as well.

Folks around the Bay of Quinte were intensely active. According to the Powles Report, "Between February 9 and April 2, 11 petitions were presented before the House of Commons, most from the Quinte area, but some from as far as Oshawa."[9]

Mackenzie Bowell

At this time, Charles Tupper was the minister of Public Works as well as the minister of the brand-new Department of Railroads and Canals. During the budget debates in February of 1880, he made a speech in which he responded to the intense lobbying for a canal in a very careful way, as any good politician should. He first agreed with the advocates about the need for the canal and was eloquent in touching on the benefits that had been raised in the recent reports.

However, his final thought was that the government would not be able to allocate significant funds for the canal in this budget year, due to many more pressing needs. Of course, he was alluding to the trans-continental railway, which was the highest priority of his government at that time. Advocates for the canal were grateful for these careful words, which did not exclude the allocation of funds from next year's budget.[10]

Charles Tupper

In the civil service, the momentum continued. Another survey was ordered early in the summer of 1880, by engineer G. E. Austin, who reported to H. T. Perley, chief engineer for the Department of Railways and Canals. This survey focused on the Presqu'ile route, in particular the situation related to rock formations toward the west end that had been reported in Rowan's survey in 1867. Test pits were dug along the route, but they did not find any rock at a depth that would cause a problem with excavation of the canal.[11]

Local business people visited with Mr. Austin while he was conducting the survey, just to make sure he was well aware of the importance of the canal in their community. They had seen so many surveys concluded with no action taken, and they wanted to make sure authorities knew that they were being watched and judged. The advocates were feeling confident but anxious.[12]

Joseph Keeler III

Then, in December of 1880, Mr. Perley advised the government that further surveys were required to determine the final route of the canal. This was good news for the advocates, but they had seen surveys result in no progress, so they kept up the pressure.[13]

Sadly, advocates for the canal lost their most forceful and effective voice when Joseph Keeler died in Ottawa on January 21, 1881.[14] Keeler had pushed hard for the canal all of his political life and was starting to see strong support for the idea. However, he was tragically denied the pleasure of seeing his pet project come to fruition.

For months, the loss tempered the debate over the canal, but it did not hamper the forward progress. Mr. Keeler had built a strong and powerful lobbying engine, and it would not be derailed by the loss of one man. Joseph Keeler's legacy is the Murray Canal.

In March of 1881, $25,000 was set aside by Parliament for the final survey, which would determine the best route for the canal.[15] Then, in April, an appropriation of $50,000 was made under the heading "Canals." The new minister of Railways and Canals, John Pope, provided the government's current position, explaining:

> There is no doubt it is a very important work, and it is more important now than ever before. It has been represented by gentlemen of undoubted integrity that if this canal were built, iron smelting works would be started at that point and a large number of men employed. Under the pressure, and after the representation that was made by those gentlemen, the Government felt they could not any longer refrain from trying to open up this important section of the country, not only for the iron but the lumber trade as well.[16]

$25,000 for the final survey

The final survey would be undertaken by Thomas Stafford Rubidge, a well-respected engineer who was, at that time, in charge of the enlargement of the St. Lawrence canals.[17] Tom Rubidge had been born in Oxford, England, in 1827. His father died when he was very young, and his mother remarried when he was nine years old. At age fourteen, he came to Canada to live with his uncle, Frederick Preston Rubidge, who was a very active civil engineer in Upper Canada.[18]

Young Tom took on the trade of his uncle and would be known as a skilled hydraulic engineer, spending most of his time in the Cornwall to Prescott area working on the St. Lawrence canals. The job he was asked to do for the Ministry of Railways and Canals regarding the Murray Canal was small compared with most of his projects, but he had extensive experience with canals that made him the obvious choice for this final survey.

Mr. Rubidge reviewed all the material from pervious surveys and, during the summer of 1881, conducted an exhaustive survey on the ground for his own immediate knowledge. His report was submitted on February 1, 1882.[19] The web site of the Library and Archives Canada provides a document from the minister of Railways and Canals reporting on Rubidge's findings and asking for "Your Excellency's approval" to go ahead with the project. A full transcription of that document can be seen in Appendix A.[20]

The most important part of this document was the recommendation for the route. Rubidge recommended a straight cut from Twelve O'Clock Point on the Bay of Quinte to the eastern point of Presqu'ile Bay. This was a slight variation on the traditional Presqu'ile route discussed in early surveys. A completely straight cut from Twelve O'Clock Point would be easier to build because it avoided the mouth of Dead Creek as well as the twists and turns of the creek as it flowed first to the west, then south and then west again.

Certainly, this route was much longer than the others, but Rubidge felt that the cost of excavation would be moderated by the fact that most of the length would be through sandy and marshy soil. He also thought that the problem of sand bars accumulating around the entrance to Wellers Bay would make navigation into Wellers Bay unpredictable and subject to bad storms that pushed the sand into the bay from Lake Ontario. This turned out to be exactly what happened. According to Marc Seguin, in his book *For Want of a Lighthouse: Building the Lighthouses of Eastern Lake Ontario 1828–1914*, "by 1900, shipments out of Wellers Bay had dropped to a trickle. The rear range light, which had already been moved twice since 1876 in order to compensate for the shifting sands at the entrance channel from Lake Ontario into Wellers Bay, was discontinued in 1909 and demolished."[21]

The shorter route to the north shore of Wellers Bay was not feasible, according to Rubidge, due to the presence of significant rock formations along the north shore of the bay, which would make excavation difficult and expensive. He addressed the issue of fluctuating water levels between the Bay of Quinte and Presqu'ile Bay by characterizing it as a rare occurrence. His suggestion was that any current resulting from this type of fluctuation would not be an impediment to traffic.[22]

The second important part of Rubidge's report was estimates of total cost for the canal based on two different routes and three different dimensions.

Width	Presqu'ile Bay	Wellers Bay
150	$974,000	$1,472,000
100	792,000	1,302,000
80	721,000	1,229,000

Estimated cost for three sizes and two routes

Rubidge may have preferred one or the other of these options, but his work as a civil engineer required him to collect all the pertinent data and present the information in a way that allowed the politicians to understand the range of options they had. It should also be noted that the document states the calculations are "based on a Scheme for a Canal of a depth of 11 feet at lowest water." In fact, this document says nothing of a depth of 14 feet.[23]

The document also shows that the minister of Railways and Canals recommended to the government the smallest of the three options presented: eighty feet at the bottom. It has been said that politicians will always choose the least expensive solutions, and this choice seems to reflect that.

I would love to see documents from the time that explain why and how this decision was made. In the absence of that, we must speculate based on what we know or expect to be true. There is a line in the Powles Report that suggests that the government, meaning the minister of Railroads and Canals, understood that the Murray Canal was not going to play a role in the wider St. Lawrence Transportation system. Despite lots of language to the contrary throughout the process, the government acted on the reality that the Murray Canal was being built for local use due to the intense lobbying from industrial and commercial interests around the Bay of Quinte.

The route decided; $200,000 for a canal

In the end, the Murray Canal was built as a reward for politicians in the Bay of Quinte area who had supported the Conservative Party in the past. The future was also taken care of by providing the local industrial and commercial concerns with the transportation link they'd lobbied for so intensely.

James S. McCuaig

It brings to mind the old saying "all politics is local." Many folks in Prince Edward County lobbied hard for the Wellers Bay route and, when the Presqu'ile route was approved by a Conservative government, they were not pleased. Of course, there would be accusations of political chicanery.

James Simeon McCuaig was a long-time businessman from Picton who was a member of Macdonald's government from 1878. However, enough people were angry about the route selection that they voted their Conservative member out of office in the next election.[24]

John Milton Platt

John Milton Platt became the Liberal member of Parliament for Prince Edward County in June 1882. He was born in Athol Township and was a doctor in Picton. As a matter of political expediency, he continued to repeat the theories about political interference that his predecessor had espoused. This was a political talking point that would be dredged up and flung about for years after the canal was built.[25]

In spite of all the political messiness, we can celebrate the work of James L. Biggar and Joseph Keeler III over several decades in support of the Murray Canal. These local politicians: Biggar from Carrying Place and Keeler from Colborne, did the grunt work in their constituencies and in the halls of Parliament that led to the final decision of the federal government to fund the building of the canal.

CHAPTER 6
Contracts and Opening the Works

Tenders for the contract for the building of the Murray Canal were opened on June 22, 1882. The contract was awarded to the firm of J. D. Silcox and Co. of Welland, Ontario, which, along with partners, Nathan S. Gere and H. J. Mowry of Syracuse, had worked on the Welland Canal, so they had lots of practical experience.[1]

The main contract was to include "the excavation of the canal proper, along with the dredging of the entrance channels and erection of the four bridges over the cut." The canal was to be eighty feet wide at the bottom and eleven feet deep.[2]

The final bid was $1,140,625, much higher than Rubidge's estimate for this size of canal. "According to Rubidge, this difference stemmed from the fact that the contract called for a much greater amount of stone protection for the canal banks than he had allowed for in his initial estimate."[3]

The contract included many aspects of canal construction that garnered less public discussion. Silcox would be responsible for building the piers and forming the sides of the canal as well as drains, side ditches, and tow paths along the banks. They would also have to do considerable dredging at the eastern and western entrances to the canal. There was about three-quarters of a mile of channel to be prepared into the Bay of Quinte at the east end, but the west end was quite another matter.

There would be about one and a half miles of channel in Presqu'ile Bay that Silcox would have to dredge. The channel would go west from the canal into Presqu'ile Bay and then turn south toward the entrance to the bay near Salt Point. Major dredging work would be required to deepen and widen the channel through the limestone shelf of the Middle Ground Shoal, east of the lighthouse. This channel had

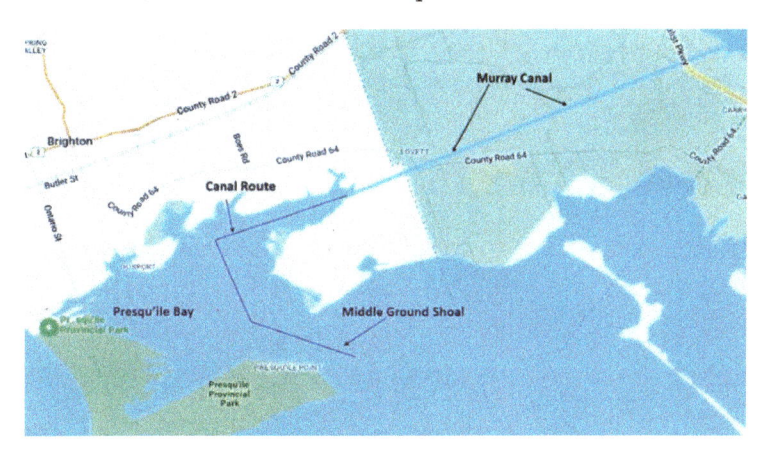

The Murray Canal — Bay of Quinte to Lake Ontario

been dredged by the federal government in 1872 but had to be deepened and widened in order to act as part of the western entrance to the Murray Canal. With the east and west entrances included, the canal would cover about nine and a half miles.[4]

There would be four bridges over the Murray Canal. One would be a railroad bridge, which would carry the tracks of the Central Ontario Railway over the canal. Besides this, there would be three road bridges. The Trenton Road would cross the canal near its east end. The Smithfield Road bridge would cross the canal near the middle, and the Brighton Road bridge toward the west end. Silcox's contract included the construction of the piers for all four bridges. The piers were to be made of masonry connected by timber crib work pilings. The same method would be used for the 500-foot-long entrance piers at each end of the canal.

Railway Swing Bridge resting on its pier, 2022

But there was a problem. "Work on the canal was barely underway when it was found necessary to alter the original bridge plans."[5] The swing bridge plans called for two channels, both fifty-four feet wide, separated by the center pivot pier of the swing bridge. However, lumber companies that expected to utilize the canal saw a problem. Their barges were fifty-two feet wide and 270 feet long. This was thought to be too tight a fit.

Several of the major lumber companies around the Bay of Quinte, led by a Peterborough lumber merchant, petitioned the Department of Railways and Canals to enlarge the width of the bridge openings. The lumber interests had lobbied long and hard for the canal and it was expected to be a boon to their business, so they did not want a small technical problem like this to impede their actual use of the canal. They made a good case that the fifty-four-foot-wide channel would not be good for them.

Very quickly, Mr. Rubidge studied the issue, and in the fall of 1882, the plan was changed. The solution was simple. The center piers would be off-set by a few feet to create a sixty-foot channel on the south side. The lumber companies were happy since their barges would now fit comfortably, and in fact, lumber would be one of the most common products carried on the canal.

Before construction could begin, a very important step had to be taken — expropriation of land. This process began with the approval on August 24, 1882, of two men to act as "land valuators," and they set to work right away.[6]

The Murray Canal would be cut from the Bay of Quinte near Twelve O'Clock Point and run across farmland, past the Trenton Road and the railway line from the county. It would continue through parts of the Dead Creek marshes, past the Smithfield Road and then the Brighton Road, to meet the eastern point of Presqu'ile Bay.

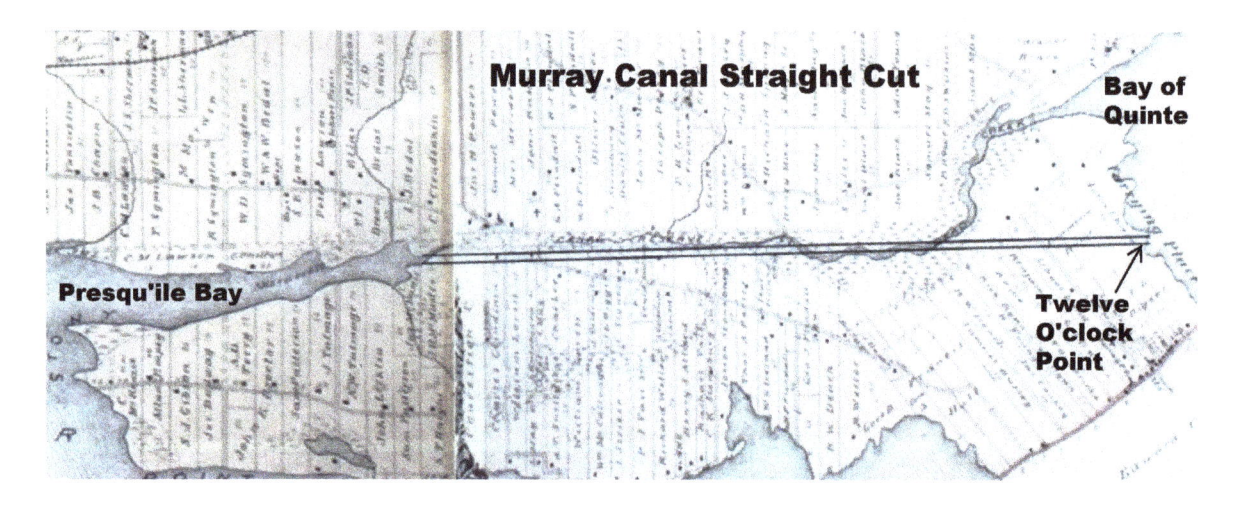

The Murray Canal cut from the Bay of Quinte to Presqu'ile Bay.

The government had the power to expropriate pieces of land from the various lots along the length of the route, and the owners had to comply. This complex process got underway in the fall of 1882 and would be completed in about six months.

Supporting the process of land expropriation were the surveyors who had to tramp over the ground and document every piece of land that was required for the Murray Canal.

An interesting document has recently come to light that tells of a reunion of surveyors that worked on the Murray Canal at the Clarendon Hotel in Brighton in 1938. Not all of them were alive by that time or able to attend, but a good contingent of the crew that was led by Adam Clarke Webb of Brighton spent some time reminiscing.

Adam C. Webb was a Dominion Land Surveyor who led the team of surveyors along the Murray Canal in late 1882 and during 1883, to define and document the route of the canal and support the land expropriation process.[7]

However, the reunion revelers in 1938 were focusing much more on an expedition this crew undertook in 1881, before their work on the canal route. This project took them to the Territories to lay out sections of land that would later support ranchers in Saskatchewan. For Mr. Webb, the Murray Canal would be a walk in the park compared to the job his crew had just completed in western Canada.

The official opening of the canal works was held on August 31, 1882, with the Sod Turning Ceremony at the farm of William Lovett, near the west end of the canal at the Brighton Road. While this was a convenient place for the event, Mr. Lovett may have felt it was appropriate to have the event here because he was preparing to have significant parts of his farm expropriated for the building of the canal.[8]

His father, John Lovett, had settled on Lot 21, Concession C of Murray Township in the 1820s and would acquire land in Lot 18 as well. William was now the owner of the land and he faced the prospect of having more than eight acres sliced off the north end of his farm in Lot 21[9] and almost that much in Lot 18.[10] He was paid well for it, but it would have been a wrenching experience to see parts of his family homestead ripped away. All the fine words about enhancing the economy might ring hollow to a farmer who had to give up land.

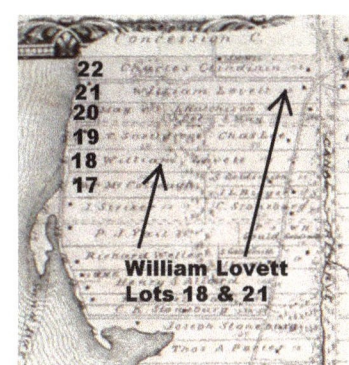

People in the area had waited for decades for the Murray Canal to finally become a reality, and there was an air of excitement as several thousand spectators gathered to listen to the bands from Brighton and Trenton as well as a long list of politicians who must give their speeches. Mayors and reeves from area towns and municipalities mingled with lumber barons and mine owners. Farmers and lawyers were part of the crowd, along with several members of the press.[11]

Opening of Canal Works at William Lovett farm

While it was a celebratory event, there was also a current of sadness to the proceedings due to the recent passing of Joseph Keeler III. The ceremonies began when Thomas Webb, the reeve of Brighton Township and chairman of the Canal Celebration Committee, presented a silver spade to Octavia Keeler, the wife of Joseph Keeler III. It would be her job to turn the first sod for the Murray Canal, in honor of the hard work and persistence of her husband in pushing for this project.

Octavia Phillips, wife of Joseph Keeler III

When the moment came, Octavia Keeler shoved the silver spade into some loose soil and deftly dropped the dirt onto one of Silcox's earth scrapers. Mackenzie Bowell added another spade full of dirt to complete the ceremony.[12]

The silver spade that Octavia Keeler used to turn the sod for the Murray Canal was then gifted to Mrs. Keeler. It later came into the possession of her grand-daughter, Ruth (Boyer) Brown[13] and was later donated to Proctor House Museum in Brighton. Here is the silver spade at the 2014 History Open House event in Brighton.

Immediately after the sod-turning ceremony was completed, there were a few moments of excitement as many more individuals wished to turn a spade of dirt into the scraper. There was much confusion and anxiety as the crowd pushed forward. Soon, however, the mass of people moved off to the platform where the speeches were about to begin.[14]

Silver spade used to turn sod for the Murray Canal

Mr. Bowell was the first speaker and he provided the appreciative audience with a brief history of the canal, saying that "it was only due to 'unbelievers and creakers [who] had thrown a hundred obstacles in the way of its construction' that the Murray Canal had not been built prior to this time."[15] Here was an experienced politician who knew exactly what kind of language would please the crowd.

His list of potential benefits for the canal seemed to come from a couple of decades before, but the crowd nodded in agreement when he said, "But now, with the North West opening up and a flood of grain expected on the waterways, the local and national benefits of this work could no longer be denied."[16] He also cited the military value of the canal but hoped that it would never be needed for such a purpose.[17]

Mackenzie Bowell

Mr. Bowell also felt it necessary to defend the decision about the route, which still raised hackles in the area. He felt that "any criticism of the final selection was unfounded, for the government had a duty to the country to adopt the route which presented the least engineering difficulties and which was recommended by the engineers who had made a careful survey of all the routes which had been proposed." Mr. Rubidge would be pleased.[18]

After Bowell was finished, several others took to the platform and presented speeches that were consistent in their praise of Joseph Keeler and repeated the verbiage about the canal bringing prosperity to the region and the country. When the speeches were finished, enthusiastic cheers were expressed for the queen and the organizers of the event. By the time the event ended and the crowd dispersed, it was late in the afternoon. The Clarke House, which was near the east end of the canal route, held a banquet for invited guests before they returned to Belleville on the midnight express train.[19]

As the revelers returned home, in the darkness along the canal route several of Silcox and Mowry's large wheeled earth scrapers were sitting in anticipation of the start of work on the canal the next day.[20]

The first job was to remove the trees and brush and scrape off the topsoil along the canal route that was marked by the surveyors. This work began in the fall of 1882 and progressed through the winter, completed by the spring of 1883.

For the scraping, dozens of wheeled scrapers were used, such as this one at the Ameliasburgh Museum. Local farmers were hired with their teams of horses to drag these scrapers, providing just one example of the many ways that the building of the canal inserted cash into the local economy.[21]

The expropriation of land along the canal route was completed by the spring of 1883. See Chapter 14 for details about land expropriation.

Horse-drawn wheeled scraper

CHAPTER 7
Building the Canal

Dredging in Presqu'ile Bay started September 1 and continued until winter forced the end the season. The barges wintered in the harbor near the west end of the canal route. These dredges were large mechanical systems floating on barges and powered by steam engines. It was slow, methodical work with complex machinery and lots of problems, but these folks were experts.[1]

The Murray Canal would require four bridges in total. The railway bridge would take the tracks of the Central Ontario Railway across the canal. The Prince Edward County Railway had begun operations in late 1879, and, in early 1882, became the Central Ontario Railway. The track ran from Trenton, across a bridge near the mouth of Dead Creek, and south for a short distance just west of the Trenton Road. Then, the line went south to approach the eastern shore of Wellers Bay near the west end of the old Portage Road, where there was a flag stop. It then continued across Prince Edward County to Picton. There is a good history of this railway in the book called *Desperate Venture*.[2]

As stated in Chapter 6, there would be three road bridges over the canal: the Trenton Road bridge at the east end; the Smithfield Road bridge in the middle; and the Brighton Road bridge at the west end. The contract with Silcox covered the construction of the piers for the bridges as well as the entrance piers at each end of the canal.

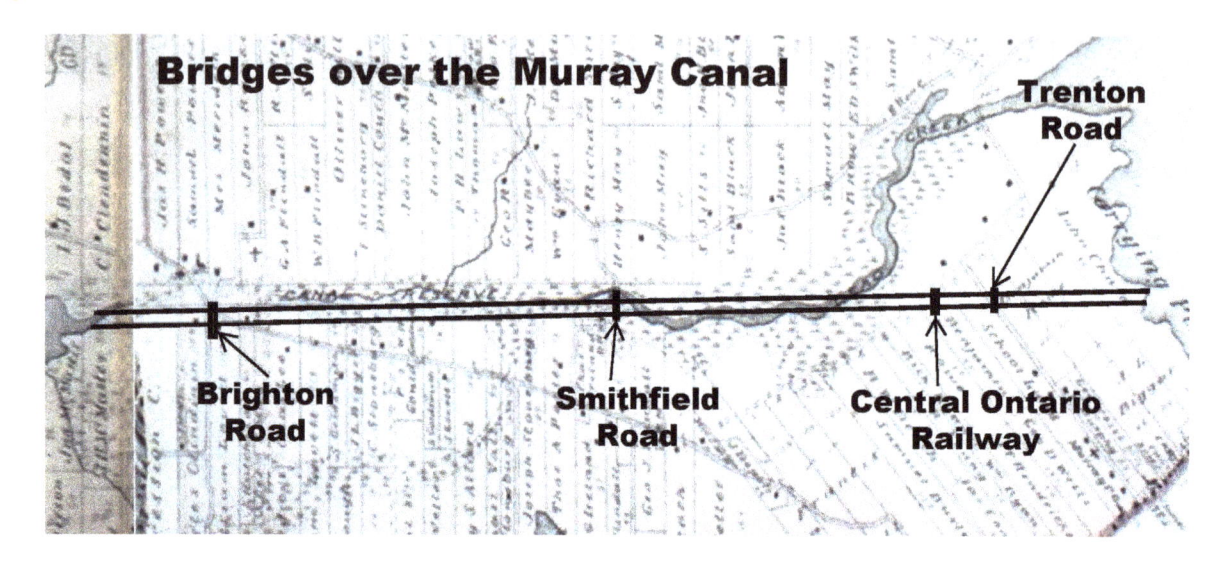

One railway bridge and three road bridges across the canal

The first full year of the canal works in 1883 saw six of the Silcox and Mowry steam dredges working constantly from spring to fall. Three of them labored in Presqu'ile Bay, creating the channel at the western entrance to the canal and on into the bay as well as enlarging the channel through the limestone of the Middle Ground Shoal east of the lighthouse. Two dredges were working to create the channel from the east end of the canal into the Bay of Quinte. The sixth dredge operated in the Dead Creek Marsh.[3]

The foundations for the piers that would extend the canal at each end were laid during 1883. They were "sunk into position one foot below the canal bottom, in readiness for the cribs."[4]

Work on the actual cut of the canal was commenced from each end, working toward the middle. By the end of the 1883 season, they had cut almost 2,000 feet of the canal at each end.[5]

This year also saw the commencement of work on the bridge substructures. The map shows all four bridges that would cross the Murray Canal. The Powles Report says that "during the summer the foundations of the piers and abutments for the Trenton Road bridge were laid. These consisted of pine timber crib work, 12 inches square, which was laid transversely nine inches apart and the spaces between filled with liquid concrete."[6]

The bridge taking the Trenton Road over the canal was considered to be critical for local road traffic and work on the underlying structures was pressed forward quickly. The next phase of the work was the masonry over the substructures, and that began by late September.

One of the many civil engineers who took part in the building of the Murray Canal was destined for much greater things. John Laing Weller was the youngest son of the old Stage Coach King, William Weller of Cobourg. Young John had graduated from the Royal Military College in Kingston in 1880 with a first-class certificate in engineering. By around 1884, he was living in Brighton and working on the Murray Canal. He would later be superintendent of the Welland Canal and oversaw the construction of the next generation of that canal in the early 1900s.[7]

John Laing Weller

The year 1884 saw the canal works intensify with good progress in all areas. Cribs for the breakwater piers at either end of the canal were formed and prepared so that they could be sunk into position during the next summer. Masonry work on the Trenton Bridge substructure was completed to the waterline, and the crib work connecting the piers was finished. Ballast for the cribs was provided by 1,240 cubic yards of stone brought in from the Point Anne quarries to the east.[8]

The Powles Report describes an interesting development in 1884. "From this point onwards, the ongoing excavation of the cut appears to have been done entirely by dredging. This is somewhat curious, for in most works of this kind the ends of the cut are left sealed and the excavation work done in dry earth. In this case however, it seems that the excavation was taken to a certain depth by earth scrapers and then the water let in and the remainder done by dredging."[9]

This approach caused some discussion at the time, but the objective may be more economical than technical. Dredges were highly mechanized devices that required a small number of skilled men to operate. Wages were still very low from a modern perspective but were increasing in the 1880s, causing operators to use technology instead of manual labor where possible. Silcox and Mowry were very experienced at this type of work, and the dredge crews were highly efficient. It was seen as a method of keeping costs down for the overall project, and for the contractor, it may have been a way to complete his contract within budget.[10]

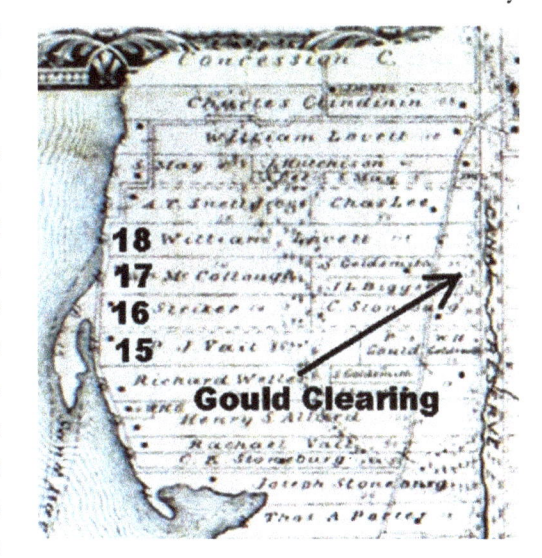

Gould Clearing in Dead Creek Marsh

In order to use the dredges in this way, dams were created to allow water into a certain defined area where the dredge was able to function without interfering with the work near the bridges where piers were being built. During 1884, two dredges were working in this manner in the Dead Creek Marsh and two farther west where the marsh narrows, called Gould Clearing. Work at the Bay of Quinte entrance continued with one dredge, and two were employed in Presqu'ile Bay to work on the navigation channel through the bay as well as the task of expanding the channel through the Middle Ground Shoal.[11]

Gould Clearing is not a well-known name today, but it probably refers to the land at the north end of Lot 17, Concession C, on or south of the Canal Reserve. This is the lot immediately east of William Lovett's farm where the Brighton Road crossed the canal. The original Crown Land Patent had been acquired by Joseph Gould in 1821, and some of it was in the hands of his descendants well into the 1900s.[12]

Even though the Murray Canal was built in the 1880s, we are lucky enough to have a few pictures that were taken during construction.

Here is a dredge working at the eastern entrance to the canal near the Bay of Quinte. From the fall of 1882 to 1886, there were as many as six dredge machines working at various location across the canal route.

The picture shows a swing bridge pier in the middle of the canal and a canal wall to the right. In fact, we are looking into one of those sixty-foot channels which were created after a petition from the lumber interests resulted in a design change to accommodate their fifty-two-foot barges.[13]

We can go a step further and show some of the pictures we have of dredge machines. Here are images of the type of dredge that would have been used during the construction of the Murray Canal and afterward for maintaining the canal.

However, all of these images show the sign of Robert Weddell of Trenton. This suggests that the pictures are from a later period, probably in the 1900s or 1910s when Robert Weddell Jr. was doing repair and upgrades at the entrances to the canal or working on harbors at Trenton, Belleville, or other places in the area.

We must remember that the contract to build the Murray Canal was given to J. D. Silcox and Co. of Welland, Ontario. This was an experienced and well-resourced company who would likely have brought their own dredging equipment to the Murray Canal.

In the absence of documentation to the contrary, it appears as if the dredging of the Murray Canal was done by Silcox equipment and crews and not by Robert Weddell's company. His contract focused on the road bridges, which was his primary business in the 1880s.

We know he had at least one dredge in 1890,[14] but no evidence is available to indicate any of his dredge machines worked on the canal in the 1880s. More documents from that time would be useful in clarifying this question.

Robert Weddell Jr. Dredges

In March of 1885, the Silcox company requested and was granted a reduction in their deposit related to the contact from 1882.[15] The chief engineer reported that the work stipulated in the contact was about one-third done, so this reduction was a reasonable step. However, this was also in the context of the fact that the work was far behind schedule.

The completion date in the contract of July 1, 1885, came and went almost unnoticed. In fact, the work was far from complete. Very close to that July 1 date, work on the substructure for the railway bridge began.[16] This bridge was a high priority, but construction was complicated by the fact that collaboration with the railway was required in order for the trains to run while the canal was being built.

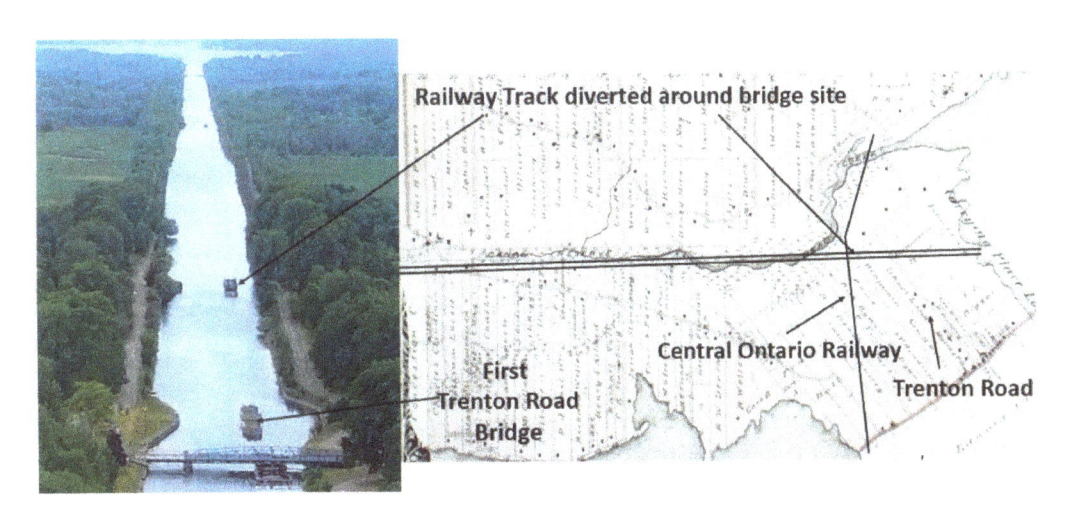

Murray Canal looking west: Highway 33 bridge, first road bridge, railway bridge

It was determined that the track must be diverted around the site of the bridge. The diversion track was completed by July of 1885, and work on the bridge substructure was given a high priority. The railway company would complain about the expense and difficulties of managing the tight turns of the diversion track, but they were able to operate more-or-less normally throughout the building of the canal.[17]

Work on the Brighton bridge was delayed because significant quantities of rock had been found in that area to the west end of the canal route. Rowan had identified "the presence of nearly 2/3 of a mile of rock which was found at a depth of about nine feet"[18] during his survey in 1867, although a survey done in 1880 by digging test pits did not find enough rock to worry about. Mr. Rubidge had done test borings and found "two or three sections of rock near the west end of the route, which would have to be cut out."[19]

By the summer of 1885, the true extent of the rock formation in the immediate area of the Brighton bridge was determined. In order to establish an eleven-foot depth for the canal over the rock, dams were built to create a dry area and work crews had to remove at least two feet of rock over a distance of about half a mile. This was not the same work being done on the Middle Ground Shoal at the entrance to Presqu'ile Bay. Operating a dredge on solid rock was much different than on sand and mud. It took a lot longer, and the cost of the work would increase far beyond estimates.

In spite of the constant activity, the canal works were far behind schedule. The Powles Report states, "It could hardly be argued that the excavation of the canal and erection of four bridges was a particularly difficult task, so what could possibly explain the delays?"[20]

A local politician felt he had the answer. Dr. John Milton Platt was the Liberal member for Prince Edward County in the federal Parliament. He reported to the House of Commons that nearly half of the dredging had to be repeated because the walls of the canal collapsed over the winter.

Mackenzie Bowell, the minister of Customs, and John Pope, the minister of Railways and Canals, dismissed this as local gossip, suggesting that the resident engineers reported no such problems. Bowell stated, "I think the work is of much greater magnitude than they supposed when they took it in hand. I am also under the impression that it is not being pushed as rapidly as contemplated — why I cannot say."[21] This is typical politician-speak for a situation where they know there is a problem, but they dare not admit it because that would put them in a bad light.

In fact, there was a problem. Thomas Rubidge had suggested that the cost of the longer cut to Presqu'ile Bay would be mitigated by the fact that the soft sand and marshy soil through much of the route would be easy to manage. That turned out to be a two-edged sword. Certainly, the first cut through soft soil was easy, but they found that parts of the canal banks that had been created during the work season collapsed during the winter. Much of the dredging had to be repeated, and much more rock had to be brought in to secure the banks of the canal. This would be a long-term problem that required constant maintenance.[22]

While the problem related to the canal banks certainly caused delays and could account for some of the extended time for the building of the canal, another factor was pointed out in the Powles Report. "An even more obvious reason for the delays was the decision to excavate the canal by dredging, instead of a dry cut. This fact was acknowledged in the Department of Railways and Canals Annual Report for 1890, when it attributed the delays to 'the fact that fully nine-tenths of the excavation was accomplished by dredging during the season of navigation, and that the work was suspended during the winter months.'"[23]

Even Mr. Powles could not establish the contemporary rationale for the decision to dig the canal in this way. It may have been due to the desire to reduce labor costs, but it certainly contributed to the delays in completing the project.[24]

In spite of all the delays, six dredges worked from the beginning of the season in 1886 with the objective of finishing the cut in that year. In fact, the final cut was made early in the summer and the waters of Presqu'ile Bay met the waters of the Bay of Quinte for the first time. There was no official celebration, but the event was identified with the shriek of steam whistles on the dredges along the canal route. For the workmen, it was a satisfying moment.[25]

During 1886, any road traffic over the canal was still confined to temporary bridges that had been erected by the contractors. Finally, on July 25, 1886, tenders were let for the building of the superstructures of the bridges and the contracts were awarded in August. The Dominion Bridge Company of Montreal received the contract to build the steel railway bridge, based on their low bid of $11,157.[26]

The same firm submitted a bid to build three steel road bridges at a cost of $5,900 each. However, to the surprise of many, Robert Weddell's Trenton Bridge and Engine Company was awarded a contract to build "three composite timber and iron structures at $6,739 each." The minister of Railways and Canals, John Page, provided an explanation for his decision, as seen in the Powles Report. He said that "a timber structure would be more durable and could be repaired more easily by local work men."[27]

There was some scepticism about this because everyone knew of Weddell's close association with the Conservative Party in the area. This idea gains more credence by the words of Edward Cochrane,[28] the Conservative member for Northumberland East, in a letter to John Pope. He said, "Anything that can be done to give R. Weddell of Trenton the contract for the superstructures on the bridges on the Murray Canal will be a great benefit to our party in the Riding as he is a man that will do anything that he can to assist us." This seems to have moved the process along quickly, and the contract was signed September 17, 1886.[29]

In fact, the awarding of this contract to Robert Weddell's company should not have been a big surprise. Robert Weddell Sr. (1821–1898) had moved his family from Edinburgh, Scotland, to Trenton in 1873, where he quickly created the company that would build many of the bridges that crossed rivers, creeks, railways, and canals in the area.[30]

His son, Robert Weddell Jr. (1850–1919), was twenty-three when the family moved to Trenton. He and his father worked in partnership until Robert Weddell Sr. died in 1898. In some documents, it may be unclear exactly which man is being referred to during the building of the Murray Canal. I am assuming it is Robert Weddell Sr.[31]

The Weddell men, father and son, were aggressive businessmen who knew how to obtain large government contracts. Robert Sr. quickly became involved with John A. Macdonald's Conservative Party and was a member of the Masonic Lodge in Trenton.[32] Although there may have been some question about the awarding of the contract for the three road bridges, there was little complaint about the work itself, which was undertaken quickly and with good results.

By the end of the 1880s, the Weddell company could boast a long résumé of successful projects and an impressive client list. An article in the *Belleville Daily Intelligencer* on August 23, 1888, provides some details:

Bridges by Trenton Bridge

The Weddell Bridge and Machine Works is one of the flourishing institutions of Trenton. Starting some 12 years ago, it was burned down once. Since rebuilding it has quadrupled its capacity for work, the latest ordered tool being a steam hammer. Its first important start was in connection with the Central Ontario Railway and Iron mines and then the Murray Canal and Bridge building. The proprietor, Mr. Robert Weddell, employs inside the works, between 30 and 40 hands who have to work over time to keep up with the orders under contract, and a large gang of men outside putting up finished work. He pays out from $1,200 to $1,500 a month and has turned out the last three months between thirty and forty thousand dollars' worth of work. The orders on hand will tax their energies till next fall. Besides general machinery, millers' contracts and repairing they have constructed the following bridges:

Three composite bridges on the Murray Canal for the Dominion Gov., three on the Trent Valley Canal, one at Carlton Place 12 ft. wide, 240 ft. span, one of the finest in the Dominion; two on Little Nation River 150 ft. span steel bridges, roadway and sidewalk. A steel bridge at Marmora 125 ft. span, and have a contract for a bridge at Port Robertson 204 ft. span, 4000 ft. of piling, structure and superstructure.[33]

All of this had been accomplished before 1888, under the management of Robert Weddell Sr. After that, Robert Weddell Jr. would carry on and expand into other profitable work.

CHAPTER 8
Demonstration and More Work

While the actual meeting of the waters occurred earlier in the summer of 1886 without much fanfare, a proper celebration was planned for the fall. This poster was made public on September 27, 1886, advertising the "Murray Canal Demonstration" to be held on Wednesday, October 6, 1886. This was called a demonstration for marketing purposes but was also described as a "preliminary opening." It would provide lucky participants with a sail through the new canal. Several craft had already made the trip, but this was to be a very choreographed event, targeting politicians and dignitaries.[1]

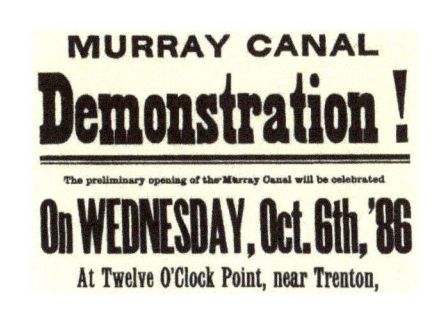

Poster for the Murray Canal Demonstration

In fact, there was still a lot of work to be done across the canal, and participants in the demonstration would see clear evidence of that, with machines and mud from one end to the other.

One of the politicians who attended the demonstration was none other than the prime minister of Canada, Sir John A. Macdonald. He was accompanied by Mackenzie Bowell, the minister of Customs and John White, the minister of the Interior. These gentlemen, accompanied by others, took the train to Brighton and proceeded from the Brighton train station to the wharf on the north shore of Presqu'ile Bay, off Harbour Street. Here they boarded a steamer, which made the trip across the bay and through the canal to Twelve O'Clock Point, arriving right on time for a substantial dinner.[2]

After dinner, the speeches began. A large crowd gathered around to listen. However, the event did not proceed as expected. The Powles Report provides an eye-witness account of what happened:

> A platform had been erected for the speakers, but it was of such insecure nature that it would not stand the strain of the people who sought to crowd upon it. Hardly had the meeting convened, when there was an ominous cracking of timber, and suddenly, I saw that portion of the platform on which the speaker's table was standing sink out of sight, carrying with it Sir John A. Macdonald and a dozen other people. The reporters' table, which was on the right of the platform, retained its position, although one or two of my colleagues also slid into the gap. Fortunately, the drop was only about five feet and none of those who went in the hole were hurt, although everybody was alarmed for the safety of the Old Chieftain. When he was pulled out from among the fallen timbers, we could see he was all right.
>
> Standing on the steps leading to the platform, the Old Man, pointing his finger to the audience, had to have his little joke. He remarked, "I have been in worse holes than that. This indicates to you folks the strength of the [L]iberal platform. It takes more than a Grit to keep me down!"[3]

In February of 1887, a mundane financial transaction took place that would go unnoticed by the men working on the canal and the local residents waiting patiently for its completion. J. D. Silcox and Co. had paid the government a five percent security deposit of $42,000 on the signing of the contract in 1882. In March of 1885, one half of the security deposit was returned to the company in recognition of progress in the project.[4] The remaining deposit amounted to $21,000.

Now, Silcox requested that the remaining deposit be returned. The chief engineer of the Government of Canada determined that "ample security rests with the Government for the due completion of the works, which [were] he state[d] proceeding fairly and that the balance of the five per cent security might as desired be released."[5]

This may be a predictable transaction based on terms of the contract, but it does indicate that, by the winter of 1886–1887, the government was satisfied with the progress made on the building of the Murray Canal. At least, there were no major disputes that would prevent completion of the financial and legal steps included in the contract.

Work at the Trenton bridge was well along later in 1886. Robert Weddell's company worked quickly to complete the superstructure and approaches by the spring of 1887. Over at the Smithfield bridge, work on the superstructure and masonry was well along at the end of the 1886 season and completed early in 1887.[6]

The railroad bridge was quite another matter. The Powles Report states, "Dominion Bridge's contract was awarded in October, but there were several delays in the establishment of their plant. Apparently, they were hampered by a lack of capital, as indicated by their continuous requests to the Department for additional funds. This was somewhat surprising for such a large firm, but by May of 1887, no progress had been completed aside from the delivery of metalwork and timber to the site."[7]

This one delay caused other problems along the canal. E. H. Laroche was a subcontractor to Silcox, who was responsible for hauling stone from Point Anne to the Smithfield Road site. The unfinished nature of the railroad bridge caused Laroche's stone barges to sit idle at the east end of the canal, prevented from moving through the canal, such as it was, to the Smithfield site.[8]

Even more of a problem for the railroad was the inconvenience created by the diversion of the track around the bridge site. While they were able to manage the diversion for the most part, it caused serious delays in train traffic. Several accidents occurred near the switch to the diversion track with over $3,000 in damages.[9]

In winter, railway workers had to clear snow off the diversion track by hand because the snow plow could not maintain the required speed around the tight curves. Officials of the Central Ontario Railway complained constantly about these problems, but the government continued to assure them that everything possible was being done to speed up the process. In fact, this would remain a significant problem well into 1887.[10]

In spite of these difficulties, the Smithfield Road bridge was completed during the spring of 1887 and the railway bridge in October. The Brighton Road bridge was still being delayed by the work to clear the rock in that area and work continued on the entrance to Presqu'ile Bay. The piers at the entrances to the canal were almost finished by the end of the 1887 season, to be completed the next spring.[11]

According to the Powles Report, "these were composed of 25' by 30' timber cribs, which were located 24 feet apart and connected by stringers at nine feet over the low water line. Filled with ballast, they extended 500 feet beyond the canal entrances at both ends. They served not only as breakwaters, but also prevented the build up of silt in the canal entrances."[12]

The year 1887 saw the inception of another major infrastructure project on the Bay of Quinte, one that would be seen by local residents to work in collaboration with the Murray Canal to facilitate trade and bring prosperity.

Several well-known businessmen of the area, including Henry Corby of Belleville, came together to form a company that would construct a bridge over the Bay of Quinte from Belleville to Rednersville. The Bay of Quinte Bridge Company was privately funded, aided by generous subscriptions from the public. Work would begin in 1888.[13]

Bridge over the Bay of Quinte at Belleville

CHAPTER 9
"Canalis" Tours the Works in 1888

A correspondent in Trenton working under the pen name "Canalis" took a tour of the canal works in August of 1888 and provided their impressions from one end of the works to the other. The article appeared in the *Belleville Daily Intelligencer* on August 23, 1888, and the text of the article is available in Appendix B.

"Canalis" article on the Murray Canal, August 1888

The timing of this tour was before the canal opened for navigation in 1889 and after most of the dredging was done. The canal cut was complete in August 1888, except for a distance near the Brighton Road at the west end of the canal, where crews were still cutting rock down to the level needed for a consistent depth of the entire canal. Lots of last-minute work was being done in preparing the canal banks, which included applying "rip-rap" stone and establishing the tow paths.

Canalis was able to speak with Mr. Mowry, one of the contractors, although they did not interview Mr. Silcox, who had taken ill. Near the west end of the canal, where the contractors' office was located, the foreman of the works, Mr. McAulicliff, spent some time describing the details of the work and the problems encountered. The question of when they would be finished was top-of-mind for everyone.

The article begins with "'Canalis' of Trenton contributes to the Empire a highly interesting article on the Murray Canal, which is rapidly approaching completion."[1]

The tour started at the east end of the canal, just in from the Bay of Quinte. The correspondent then, commencing at the eastern end, gave the following description of the work:

> The cutting here is through a fine sand, like the ordinary beach sand, but further up the work I had heard that rock was met with. The banks are here "rip-rapped" with stone to protect them from the swell caused by vessels passing through and the spring floods.[2]

The correspondent then reached the Trenton Road bridge and the railway bridge. It's quite possible they were walking on the bank of the canal rather than riding in a boat on the waters of the canal. Observing the road bridge provided an opportunity to praise a fellow Trentonian:

> The road bridges, two of which are completed, swing on a huge central pier protected on either side by other piers, and are built by Mr. Robert Weddell, Jr., at his bridge works at Trenton; and bear out in their beauty and strength the great reputation which Mr.

> Weddell has earned by the quality of his work of being one of the foremost bridge builders in Canada.[3]

Canalis touched on the idea espoused by opponents to the final route of the canal that quicksand in the Dead Creek Marsh would cause machines to disappear into the muck:

> Further up the cutting is through a large marsh, called "Dead Creek Marsh," where it was supposed to be impossible to find solid bottom, and where the opponents of the present route prophesied the abandonment of the work on account of the quick-sands. But here, strange to relate, so far from being troubled with quick-sand, the banks are the best on the whole length of the cutting.[4]

Farther along the canal, Canalis passed the Smithfield Road bridge and approached the place where the Brighton Road would cross the canal, at William Lovett's farm. Here was the work that had caused delays in the canal works and prevented the third road bridge from being completed:

> Passing the second road bridge a dam across the canal is arrived at, and another can be seen a short distance farther up. Between these dams, the water having been drawn off, appears the only rock met with in the whole route of about five miles. Had the contractors been aware of the existence of a bed of rock at this part of the cutting of such a size and of a quality equal to at least to any used on the works, they would have been saved the expense of procuring stone from the Point Ann quarries for the piers of the bridges. This rock is very irregular in its formation; rising to an elevation of about eight feet from the bottom on the southerly side, it gradually diminishes in thickness, in the form of steps towards the north, until it disappears altogether, or leaves only a thin ledge to be removed in order to obtain the required depth of the excavation. In this form the rock extends for about 2,800 feet, and was a complete surprise to the engineers, its existence being unknown until struck by the dredges.[5]

The description of the dams, which provide a dry area for working on the rock, matches the picture we have of this site. More important, this is a very interesting first-hand description of the work being done to remove the rock formation at the west end of the canal. However, it seems odd, from our vantage point today, that the correspondent had made such a strong point about how surprised the engineers were to find this rock formation at this location.

Here, we see a story of local lore developing, with the intent of molding the problem that delayed the opening of the canal into some kind of mystery. In fact, the various surveys done over

Dry work to remove rock west of Brighton Bridge

decades leading up to the actual building of the canal reveal that this rock formation was detected and documented in the Rowan survey of 1867.

The Presqu'ile route was selected because Presqu'ile Bay made for the best western terminus of the canal. The cost of clearing some rock toward the west end of the canal was included in the estimates. It was no mystery.

The objective of the correspondent at this point was to interview the foreman of the works, Mr. McAulicliff, who was working at the office of the contractors, just north of the Brighton Road bridge:

> A short distance north from the third bridge site is a post office and quite a little village, where before the commencement of the canal there was only a solitary farm house. The buildings are principally boarding-houses put up for the accommodation of the workmen. The headquarters and office of the contractors being at this point of the works I proceed thither in quest of the foreman of the work, Mr. McAulicliff.[6]

Murray Canal road sign on County Road 64

The modern driver will see the sign on County Road 64 approaching the swing bridge, indicating the village of "Murray Canal." On the other hand, looking at Google Maps, that name is not evident, but the name "Lovett" is printed just south of the bridge.

The legacy of William Lovett's homestead and farm is perpetuated in the name "Lovett," but the road signs are meant to support travelers, so "Murray Canal" it is.

The foreman and the correspondent proceeded west along the bank until they came to the piers at the western entrance to the canal:

The name Lovett on the map

> Arriving at the end of the bank, we found two lines of piers extending out into Presqu'Isle harbour similar to those found at the entrance of the canal from the Bay of Quinte. I learned, on questioning my comrade, that these piers which are almost completely filled with stone, were to be planked over with heavy plank, every other one being fitted with a tie post, so forming for all practical purposes, a long dock on each side of the canal.[7]

The channel here was dredged to a width of 200 feet and a depth of sixteen feet. The foreman was proud to point out that farther out, where the channel meets Lake Ontario, the channel is 1,000 feet wide.

Looking out into Presqu'ile Bay from the western pier, Canalis could see "the lighthouses and the white tents of the campers on Presqu'Isle Point, which is a noted summer resort for the surrounding townspeople."[8]

By the late 1880s, camping on Presqu'ile Point had become a very popular summer activity, especially for the growing middle class in Brighton and area. The Murray Canal would bring more visitors, which spurred the development of the hotel and dance hall in the early 1900s.

Of course, the correspondent could not pass up the opportunity to recognize Brighton: "Northwest about a mile from the north shore of the bay, among the trees, appears the picturesque little village of Brighton, with its pretty houses and wide clean streets."[9]

On the question of completion time for the canal works, Canalis says:

> In conversation with Mr. Mowry, one of the contractors, as to the time of completion, I was told that the whole of the clay excavation and the trimming up of the banks would be completed this fall. Mr. Mowry said that he did not think it possible that all the work between the dams would be removed by that time; but since the winter would not hinder in the least the work of blasting, it would be all removed by the spring, leaving then only the dams to be cut away to render the canal open to navigation.[10]

This was good news for everyone awaiting the full operation of the Murray Canal. At the end of the article, the correspondent discusses the benefits of the canal:

> These are, indeed, the chief apparent advantages, but any one for a moment supposing it would turn the course of the great highway of Canadian commerce on the great lakes in that direction is in error. Our commerce there is no longer carried on by means of canoes, but by vessels as large and strong as those with which our forefathers crossed the Atlantic and doubled Cape Horn. The coast of Prince Edward presents no longer any terror to the Canadian mariner. So, it will appear that the only through trade which will pass through the canal will be that carried by passenger steamers from Toronto to Montreal, calling at intermediate ports.[11]

Here is one more expression of the idea that the Murray Canal would be a local transportation link and not part of the broader Canadian trade system.

CHAPTER 10
Final Work and Opening

The Department of Railways and Canals produced an annual report, listing the projects being undertaken across the country. The 1888 report described the Murray Canal as having three road bridges, which may mean that the Brighton Road bridge was finally completed by then. It reported that all the excavation was completed. While the canal was not yet officially open to traffic, "some vessels were allowed to use the canal with the permission of the contractor." Work to be done including the placing of "rip rap" along the canal banks, which refers to a layer of small rock designed to reduce erosion.[1]

In September of 1888, the appropriation voted for the construction of the Murray Canal was running out and the work was not complete. The Minister of Railways and Canals asked the government for a further appropriation of $80,000 as recommended by the chief engineer. This was needed immediately in order to push through with the last of the work. In response, the minister of Finance recommended that a special warrant be issued in the amount of $80,000, which was signed by Sir John A. Macdonald on September 29, 1888.[2]

Another detail was addressed in October 1888 when authority was given to purchase the house that bridgetender J. McCadden had built for himself and use it as a watch house at the railway crossing of the canal.[3] Thus began the long and complex saga of employees of the canal building their own homes near their workplace, only to have them purchased by the government so they could be used to house the constantly changing population of workers on the canal over decades.

The Murray Canal was opened to traffic on April 14, 1889, although Silcox's contract was not officially completed until August.[4] During the first year, around a hundred vessels passed through the canal. Technically, it was not fully complete until the next year. People using the canal could plainly see that there was still no stone facing on canal banks for about three miles of the canal. This was curious because the stone facing was clearly part of the original Silcox contract.[5]

This lack of stone presented a serious problem because it allowed erosion to occur on the canal banks, leading to larger potential problems with bank stability. To deal with this, Silcox entered into a separate contract in May of 1890, and the stone was placed by December at a further cost of $25,000.[6]

In June of 1889, the hiring of the first superintendent of the canal was completed. Thomas Phillips Keeler[7] was a son of the beloved canal advocate Joseph Keeler III, and there appears to have been some concerted lobbying from family and friends to have the son benefit from the reputation of the father.

An interesting document exists in the archives that seems to have motivated a final decision on this matter. On his way to Ottawa, Thomas Phillips Keeler stopped in Gananoque to visit a family friend. George Taylor of Gananoque provided Mr. Keeler with a personal note for the prime minister, Sir John A. Macdonald, in Ottawa. It said:

> My Dear Sir John,
>
> The bearer, Mr. T. P. Keeler of East Northumberland, who has relatives in my town, called on me this afternoon on his way to Ottawa where he goes to see you in reference to an appointment on the Murray Canal and asked me for a letter of Introduction which I cheerfully give him & will be glad if you can meet his wishes.
>
> I Am Yours Faithfully, Geo. Taylor[8]

Just a week later, after several years of lobbying, Thomas Phillips Keeler was appointed to be the first superintendent of the Murray Canal.[9] The Powles Report says, "The entire operation of the canal was under the supervision of a Superintendent or Toil Collector, who was responsible for requisitioning supplies, collecting vessel tolls and recording water levels."[10]

The Grand Opening of the Murray Canal took place on August 16, 1889. In 1939, the *Belleville Intelligencer* printed a first-hand account of that event under their "Looking Backward" banner, providing some interesting news from fifty years before:

> Through the courtesy of Mr. H. Corby, M. P., many citizens, including Mayor W. Jeffers Diamond and the members of the city council, Mr. Thomas Ritchie, President of the Board of Trade, Col. Strong, American consul here, and many other prominent citizens, joined a number of the prominent citizens of Trenton yesterday in formally opening the Murray Canal. Arriving at the canal three boats loaded with people; (with the Trenton Band playing on one of the boats) steamed through the canal as far as Brighton Wharf where a stop was made and where lay at the wharf the yacht *Surprise*, Commodore Forbes in command, which was the first boat to pass through the canal. Later the excursionists proceeded to Presqu'ile, which in the near future will become famous as a summer resort. At present there are many campers there. It was with regret that the party sailed for home. We noticed the following prominent persons present from a distance and from Trenton: Hon. Mackenzie Bowell, Minister of Customs; Mr. Cochrane, M.P.; Mr. H. Corby, M.P.; Mr. G. W. Ostrom, M.P.P.; Mr. D: Gilmour and Mayor Morrison of Trenton. Although there were four members of Parliament present in the party, no speeches were made.[11]

The yacht *Surprise* was the first boat through the canal after the official opening. Commodore Forbes would have been very proud to man the helm of his favorite craft as it made history. The *Surprise* had been built by Forbes's friend and fellow mariner, Alexander Cuthbert, who was a renowned yacht builder and designer.

Alexander Forbes was born in Scotland in 1842, and his family came to Canada soon after, settling in Cobourg, where he became friends with Alexander Cuthbert. His father, Alexander Sr., was a long-time cooper in Cobourg, and his ability to provide prosperity for his family allowed Alexander Jr. to attend law school and become a lawyer. He never married, but he put his legal training to work as the solicitor for the Molson Bank in Trenton starting in 1881. The 1901 Census for Trenton shows him living beside David Gilmour, the lumber baron, and he was also the partner of Charles Francis for a time. He was definitely part of the upper class in Trenton.[12]

On a more official level, a significant change occurred in the structure of the company that was building the canal. In effect, Mr. Silcox and Mr. Gear retired from the partnership, leaving

Mr. Mowry the sole contractor responsible for the work.

> On a Memorandum dated 17th April 1890 from the Minister of Railways and Canals representing that under date the 24th of August 1882 a Contract was entered into with Messrs. Silcox Gear and Mowry the lowest tenders for the work of constructing a Canal between the head waters of the Bay of Quinte and Lake Ontario, known as the Murray Canal, and that immediately after the signing of the Contract Mr. Gear assigned to Mr. Mowry all his interest therein and on the 18th of July 1888, Mr. Mowry accepted the Contract interest of the remaining partners Mr. Silcox, thus remaining practically the sole Contractor and Mr. Mowry asks that he be so recognized. The Minister further represents that the Canal is so far practically completed as to be used through informally, for traffic, the total payments made amounting to $987,700 and the whole of the security having been returned, all necessary deeds and documents affecting the said assignments have been lodged with the Department of Railways and Canals. The Minister recommends that Mr. Mowry be recognized as sole Contractor for these works, the retirement of Messrs. Silcox and Gear from the Partnership and conditions of the Contract remaining as at present, but applicable to Mr. Mowry alone. The Committee submit the same for Your Excellency's approval.
>
> Signed: John A. Macdonald; Approved: May 20 1890.[13]

John David Silcox had been known as J. D. since he was a child. As the work on the Murray Canal drew to a close and his part in the contracting ended, he would leave Brighton and go back to his home in Syracuse, New York. However, he would not return empty-handed.[14]

When the contract to build the Murray Canal began in 1881, J. D. had moved to Brighton in order to be close to the works. He was forty-one years old at this time and had never been married. His engineering projects had kept him on the go, from his home in Syracuse to the Welland Canal[15] and to Collingwood in support of the ship-building business there.[16]

As things developed, he would meet a young lady in Brighton, Mary Elizabeth Platt, who was the daughter of Willett Platt, one of the most successful lumber merchants in the region. And, yes, Platt Street in Brighton was named after him.[17]

In 1888, John David Silcox and Mary Elizabeth Platt were married, and by April of 1891, they were living with her parents, along with one son, George. Another son, Joseph, would come along in a few months.[18]

The records suggest that J. D. and Mary lived at the north-west corner of Platt and Main Street in Brighton. Mary had acquired this property just after her marriage, and they would sell it in 1894 to Charles M. Sanford, who was a doctor in Brighton.[19]

Sadly, Mary Silcox died in Syracuse at the age of 38 in 1900,[20] leaving J. D. with two sons to raise. He was water inspector in the town for some time and worked at various jobs before his death in 1919, at the age of 72. The Murray Canal was J. D. Silcox's last big project.

Only a few weeks after the contract was modified, two official steps were taken by the government. First, the new Murray Canal was brought into the legal framework for operating canals in Canada under the heading of "Rules and Regulations." In the records of the Privy Council Office held by the Library and Archives Canada, we can read the document describing that step:

At the Government House at Ottawa, Friday, 16th day of May, 1890.

Present: His Excellency the Governor General. The Honorable: Sir H. L. Langevin, Mr. Bowell, Sir A. S. Caron, Mr. Carling, Mr. Costigan, Mr. Smith, Mr. Chapleau, Sir John Thompson, Mr. Foster, Mr. Abbott, Mr. Haggart, Mr. Dewdney, Mr. Colby. In Council.

His Excellency under the authority conferred upon him by Chapter 37 of the Revised Statutes intituled "An Act respecting the Departments of Railways and Canals" and by and with the advicse of the Queen's Privy Council for Canada, is pleased to Order that the Rules and Regulations for the management, maintenance, proper use and protection of the Canals of the Dominion of Canada, made and established by the Order in Council of the 26th day of October 1889 (Consolidated Orders of Council of 1889), together with any amendments thereof or additions thereto, shall be, and the same are hereby made applicable to the Murray Canal with the exception of such sections or provisions as relate specifically and only to other works named therein.

Rules and Regulations apply to the Murray Canal

Signed: Chas. C. Colby; Approved: May 18 1890; Stanley of Preston [The Governor General].[21]

Another important step was to officially set the rates and toll structure applied to the Murray Canal:

At the Government House at Ottawa, Friday, 20th day of May, 1890.

Present: His Excellency the Governor General. The Right Honorable Sir John A. Macdonald. The Honorable: Sir H. L. Langevin, Mr. Bowell, Sir A. S. Caron, Mr. Carling, Mr. Foster, Mr. Tupper, Mr. Haggart, Mr. Dewdney.

His Excellency, in pursuance of the provisions of Chapter 37 of the Revised Statutes, intituled "An Act respecting the Departments of Railways and Canals" and by and with the advise of the Queen's Privy Council for Canada, is pleased to order that the rates of toll to be imposed and collected on the Murray Canal, now open for regular traffic, shall be and the same are hereby fixed at the rate of one-eighth of those charged for passage through the St. Lawrence Canals.

Signed: John A. Macdonald; Approved: May 20 1890; Stanley of Preston (Governor General).[22]

A further step taken in 1890 was the installation of lights on the entrance piers at each end of the canal. It is interesting to note that lighthouses were built by the Marine and Fisheries Department, but the Murray Canal was the responsibility of the Department of Railways and Canals. Therefore, the engineers of the Department of Railways and Canals consulted with the chief engineer for Marine and Fisheries.

The book *For Want of a Lighthouse* provides the following description of the lights that were installed:

> A fixed red light visible 4 miles from all points of approach by water, shown from a lenticular lantern, elevated 19 feet above the water, standing on a square pyramidal open frame 30 feet from the end of the north pier at the entrance of the canal. The frame is 12 feet high above the pier and is painted brown. The light was first shown on 22nd August, 1890.[23]

The common term for these lights was "copper pier lamps," and they were purchased for fifteen dollars each.

The same book provides this about the first light keepers:

> The first keepers of these lights were Fred Fitzgerald and M. McAuliff who were hired in August, 1890, to "attend" the East Pier light and West Pier light respectively. For their services, they were paid a wage of twenty-five cents per day. After only twenty-six days on the job, McAuliff was replaced by C.A. Harries who tended the West Pier light until the close of navigation that year.[24]

CHAPTER 11
Operations and Improvements

During the spring of 1890, the Department of Railways and Canals hired permanent staff, and on May 9, the bridgetenders took over operation of the swing bridges from the contractors. This signaled the end of the construction phase and the beginning of operations of the Murray Canal.

The staff of the Murray Canal grew quickly to address all parts of the maintenance and operations of the canal. There were eight bridgetenders, who earned $1.25 to $1.50 per day. There were also two lightkeepers, who earned $0.50 per day. An administrative clerk was also hired to work with the superintendent.[1]

Maintenance across the canal was done by a force of twelve laborers, who worked under a foreman for ten months of the year. In the early years of the canal, there was limited work to do. The grass had to be cut, rip rap needed to be placed along the canal banks, and the tow paths had to be tended. It was likely that many of these jobs were "patronage positions, reserved for the local party faithful."[2]

In fact, it was not only likely, but certain. "As Allan Alyea, a maintenance man on the canal, whose father also worked there in the 1920's [*sic*] remembered, 'In them days when a change in government come you picked up your dinner pail and went home because you didn't have any choice. It was just one of them things... if the Liberals was in, why if the Conservatives were working there they was all finished and vice versa.'"[3]

The bridgetenders were paid more and enjoyed a bit more prestige, but had a more difficult job. They worked alone for twelve-hour shifts and were required to perform the physical labor involved in turning the swing bridges.

The Powles Report says, "To swing the bridge the master would manually release the locking wedges, then insert a large crank into the gearing mechanism in the centre of the bridge. He would then simply walk the heavy crank around until the vessel could pass through."[4]

Lock tender turns a handle to open the lock, similar to canal bridge system.

When traffic was light, the bridgetenders kept busy cutting the grass and minding the flower beds around their station. Orders from the superintendent said that the grounds must be well maintained in order to present a pleasing appearance to the public.

As one might expect, the workmen on the canal lived in the surrounding area and reported to the canal storehouse every morning for their work assignments.[5]

During the first official year of operation, more than 101,504 in vessel tonnage passed through the Murray Canal, a level that would remain relatively constant through the 1890s. The lumber industry accounted for at least one-third of the total freight on the canal. The product most often sent via the Murray Canal was squared timber, which was floated east through the canal in rafts, destined for Gilmour's enormous mills in Trenton or the Flint and Holden mills in Belleville. While the timber trade would decline in the early 1900s due to depletion of trees in the northern regions, the Gilmour mills experienced a bit of a boom for a few years after the canal opened. Sawn lumber was also shipped through the canal, peaking in 1892 at 2,539 tons. Much of this was shipped to the large Oswego Box factory in New York State. As timber declined, other wood products increased on the canal. Firewood and railway ties were seen on the canal a good deal, usually headed for the United States.[6]

How much iron ore was carried on the Murray Canal? The right answer is very little. Sure, during the late lobbying for the canal, politicians spoke glowingly of the prospect for ships loaded with iron ore from various mines in northern Hastings County passing through Belleville and then the Murray Canal on the way to smelters in the United States. However, this never happened. The Powles Report explains:

> Iron ore was never really a significant commodity on the canal. The Coe Hill mine near Marmora, which was to have established the area as a permanent mining base, ultimately fell far short of expectations. It was opened in the early 1880's [*sic*], but by 1888 operations were terminated due to the high sulphur content in the ore and increasing American competition. Other small mines also appeared briefly to the north of Madoc during this period, but these were quickly depleted and also produced low grade ore. Thus, by the time the canal was finally completed, the mining operations in North Hastings had all but vanished and the anticipated flood of ore from Belleville never materialised.[7]

In the first months of the Murray Canal, there was public dispute as to which mode of travel enjoyed the right-of-way on the canal. Officials of the Central Ontario Railway lobbied to have the railway bridge turned so that their train schedules could be maintained. However, the conflict between canal traffic and road traffic was more troublesome.

It was determined that canal traffic had priority over road traffic, which meant that the swing bridges would be open so that boats on the canal did not have to reduce speed.[8] Modern drivers might see this as rather odd, but we must remember that during the first couple of decades of the canal, traffic on the roads consisted of horse-drawn vehicles, people riding horses along with many pedestrians. Those noisy and smelly automobiles began to appear in the 1910s and their numbers increased quickly in the 1920s. The priority for the bridges changed to accommodate the growing and often impatient road traffic.[9]

At the same time, the traffic on the canal gradually declined. In effect, there was a practical evolution of practice that saw a change of priority. Our experience now has been that the road bridges are left closed so road traffic can go across without delay and the bridge turns only when there are boats ready to pass through. But that's not how it was to begin with.

The railway learned another lesson after many incidents of local cattle wandering onto the tracks near the bridge. This led to the demise of several cattle but also caused considerable delays for trains that had to stop until the problem was cleaned up. Eventually, the railway installed fences along the tracks on canal land.[10]

The year 1891 saw traffic on the Murray Canal increase and operations settle into a pattern. Two major events happened in that year that were on the periphery of the canal but closely connected in different ways.

First, the prime minister of Canada, Sir John A. Macdonald, passed away in Ottawa on June 6. "While he was in bed recovering from a cold, a severe stroke overtook him on the afternoon of 29

May. He never spoke again. He died a week later, in the evening of 6 June 1891."[11]

Sir John A. Macdonald had been prime minister of Canada during the building of both the trans-continental railway and the Murray Canal. Of course, these two projects are not equal in any way, but during the 1870s and 1880s, both were intertwined with the efforts of the prime minister.

The other event in 1891 was the opening of the bridge over the Bay of Quinte at Belleville. This was a huge event for the region. It was thought that this new bridge would work in conjunction with the Murray Canal to boost the economy of the area and improve the lives of the people around the bay.

An article in the *Daily Intelligencer* on July 2, 1891, expressed the positive feelings created by the opening of the bridge:

Sir John A. Macdonald

> With opening of permanent communications, by means of the Bay of Quinte Bridge, between the counties of Hastings and Prince Edward, a new era in the history of Belleville has been commenced. It is anticipated that, by means of the grand steel structure which here spans Quinte's waters, the commercial relations between Belleville and the rich Peninsular County to the south will be so broadened and extended as to conduce materially to the prosperity of all who are interested therein. And so will it be, beyond all reasonable doubt.[12]

Bay of Quinte Bridge opened July 1891.

The Murray Canal warranted a writeup in the *Canada Gazette* dated October 31, 1891. The title was "Murray Canal and Approaches, Including Aids to Navigation" and the following text provides a detailed explanation of the canal, which might be useful to sailors who use the canal:

> The Murray Canal is a straight cut (tangent) 6 ½ statute miles long between extremities of piers, 80 feet wide on the bottom, and 12 ½ feet deep below the ordinary low water level of Lake Ontario, or the zero of the Toronto gauge, joining the head of the Bay of Quinte with Presqu'ile Bay in Lake Ontario… After passing Salt Point the channel gradually widens to 1,000 feet until deep water in Lake Ontario is reached.[13]

This is one of the few places we can see that the channel dredged through Middle Ground Shoal was 1,000 feet wide. No wonder it took so long. Boaters who use this channel to enter and leave Presqu'ile Bay may take its size and shape for granted, but for the tourist standing beside Presqu'ile Point Lighthouse, the presence of this channel is harder to imagine.

Minimal construction work or maintenance was needed along the canal in the first decade. One exception was an overhaul of the turntable for the railway swing bridge in the winter of 1892 and 1893. Robert Weddell Jr. obtained the contract for this work, which was undertaken during the off-season so as to minimize the interruption in railway traffic on the line.[14]

On occasion, work in the canal area was caused by accidents, such as a minor collision in July 1893, when the steamer Magnet hit the Trenton Road Bridge. Apparently, it was not too serious because the costs associated were estimated to be $30.[15]

By the middle of 1893, the government had a balance owing to the contractor and the final payment was approved by an Order in Council dated August 15, 1893. "The total amount of such estimate being

$1,052,851.45 and that of this amount there [had] already been paid a total of $1,046,670 leaving the balance payable $6,028.79." With this last payment, the contract was fully completed.[16]

There was still confusion on the issue of why the Murray Canal was not built to the depth of fourteen feet. Around the same time that the government decided to build the canal to a depth of eleven feet, the Welland Canal had been enlarged to accommodate ships from the upper lakes that drew up to a depth of fourteen feet. Those larger ships could then go all the way to Kingston or Prescott before they had to be unloaded for passage down the St. Lawrence.

In 1885, government trade and navigation statistics showed that in the previous decade, the number of ships arriving at or departing from Kingston harbor dropped from 5,500 to 4,900. On the other hand, the average registered tonnage on those ships rose from 207 to 232 tons. In fact, some of the newer ships could carry over 400 tons. Obviously, the fourteen-foot depth of the canals became even more critical as the years went by.[17]

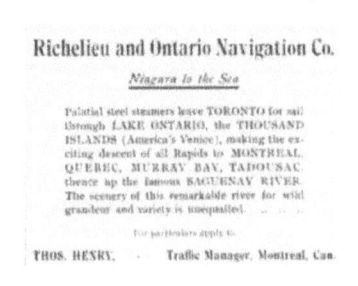

One of several companies running steamboats on the Murray Canal.

However, for most of the traffic on the Murray Canal, the depth was not an issue. "Aside from the odd schooner or sloop which had to be towed through the cut by a team of horses, the great bulk of the materials on the canal were carried by steamship. In 1895 for example, almost 158,000 tons out of a total of 162,414 on the waterway were steam vessels."[18]

Large navigation companies operated many of the steamships, such as "Royal Mail, Richelieu and Ontario, and Gildersleeve lines." There was a respectable business carrying freight and passengers between Toronto or Hamilton and Montreal.[19]

At the same time, passenger traffic made up a significant proportion of the activity on the canal. According to the Powles Report, "one of the first of these was the *Varuna*, a steamer with a 300-passenger capacity which was owned by the Quinte Navigation Company. It ran regularly scheduled trips from Brighton or Trenton to Belleville, Picton and other Bay towns, as well as pleasure cruises through the region. Just prior to World War One it was replaced by the *Brockville* of the Kingston Navigation Company, which offered chartered excursions to church groups, clubs and other organizations. These boats often had dancing on board for their passengers and would operate pleasure cruises not only in the Bay but sometimes as far as the Thousand Islands near Kingston."[20]

Steamship Varuna

Steamship Brockville

For several decades the sight of a steamboat loaded with passengers was very common on the Bay of Quinte, Presqu'ile Bay, and the Murray Canal. It represented the growing leisure time that new economic conditions provided to the public as well as the ability of entrepreneurs to produce systems and technology that the traveling public would use. Steamships were big business.

CHAPTER 12
Fixing and Changing

After a decade of operations, weaknesses were highlighted and addressed. Thomas S. Rubidge, the superintending engineer on the Murray Canal, reported in 1894 that "new and more powerful range lights [were] required at each end of the canal to indicate the dredged channel."[1] It took a while, but in 1899, William Anderson, the chief engineer of the Marine and Fisheries Department, stated:

> The Department of Railways and Canals have improved the character of the lights at the east and west entrances of the Murray Canal, adjoining the Bay of Quinte and Presqu'ile Bay. The former lights were fixed red lights shown from small lanterns standing on brown pyramidal open frames. The new lights are fixed white lights elevated 27 feet above the level of the water and visible five miles from all points of approach. The light buildings, which stand on the sites of the old frameworks, 30 feet from each end of the north pier of the canal, are enclosed hexagonal galvanized iron cabins, with cylindrical columns surmounted by the lenses rising from the apexes of the roofs. Each is 18 feet high, from the deck of the pier to the lens, and is painted white.[2]

New range lights on piers, 1889

An unusual proposal came to the Ministry of Railways and Canals from the Central Ontario Railway in May of 1899. COR wanted to build an electric railway from Twelve O'Clock Point, along the north side of the Murray Canal, up to the Trenton Road.

This diagram accompanied the Order in Council. It shows the Murray Canal with the Central Ontario Railway to the west as well as the Trenton Road. The proposed electric railway is shown hugging the north side of the canal property, from the east side of the Trenton Road, east to Twelve O'Clock Point.

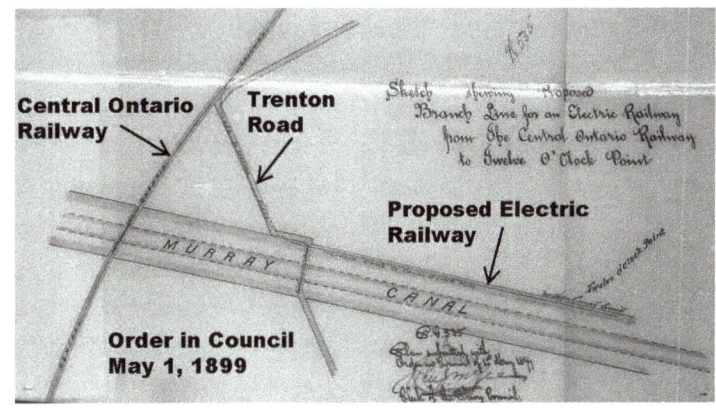

Proposed Electric Railway to Twelve O'Clock Point 1899.

The order granted permission for the COR to lease the property for the tracks for the annual rental of $50. Conditions included the assurance that COR would remove the tracks and clean up the property if they no longer wanted to lease land for the tracks. It appears as if this electric railway was never built, but it would have been a welcome facility for the patrons of the Twelve O'Clock Point resort.[3]

In early 1900, the superintendent for the Murray Canal, Thomas P. Keeler, got into hot water through some questionable financial decisions. The minister of Railways and Canals thought it serious enough to hire a law clerk to do a thorough investigation.[4] The minutes of the Privy Council dated January 15, 1900, include a memorandum asking for an appropriation of $400 to pay for this "unforeseen expense."[5]

Another memorandum dated March 23, 1900, explains that the investigation had been completed, with Mr. Keeler represented by counsel. The investigation found that there had been "abuse of his position by Mr. Keeler, to his own advantage, and to the Governments pecuniary detriment; specifically, in the undue charging for horse and vehicle in trips taken over the Canal, in the employment of Government labourers for his own domestic purposes, in the use of Government coal for the heating of his private residence, etc."[6]

Further, the minister felt that "though, perhaps, no wilfully criminal act, or intentional wrong doing [had] been committed by Mr. Keeler, his acts, nevertheless, [had] been of a character highly reprehensible in one occupying the responsible position he was filling, and that it [was] not desirable in the public interest that he should continue in the employment of the Government. The Minister accordingly recommend[ed] that Mr. Keeler be dismissed."[7]

This was an unfortunate episode in the life of the Murray Canal, although the canal continued to operate smoothly. A new super was put in place, but five years later, the position was eliminated altogether.[8]

Traffic on the Murray Canal had been steady but not spectacular since it opened in 1889. However, that would change dramatically in 1905. Stone from Point Anne was already one of the most common products carried on the canal, going west to cement and aggregate plants on the shore of Lake Ontario. The demand was growing, so "to capitalise on this resource, the Belleville Portland Cement Company opened at Point Anne in 1905, and just two years later the Lehigh Portland Cement Company was built one mile east of it. Together these firms would spark a dynamic change in canal traffic over the next several years."[9]

The increase in traffic can be seen in the numbers. Tonnage in 1905 was 29,421, and in 1907, it had increased to 52,402.[10] This kind of volume and demand for Point Anne stone products could be passed through the Murray Canal if the vessels used were kept to a limited size. As time went by, the cost-benefit analysis would tip toward the efficiency of larger ships, even though they had to go down Adolphus Reach and into the east end of Lake Ontario. In the world of ships and canals, size matters.

Wood products would decline to a trickle on the canal in the early 1900s due to depletion of the source product and the subsequent shutdown of Gilmour in Trenton, and Flint and Holden in Belleville.[11] However, in the same period, agricultural products began to pass through the canal in significant quantities. Trenton had become an important grain shipping center, and there were grain elevators on the north shore of Presqu'ile Bay at Gosport. Quantities were nowhere near what Mr. Bowell had predicted, but in 1901 alone, over 914 tons of wheat went through the canal.[12]

Apple pickers in Brighton

Apples became an important commodity at this time as well, reaching 1,349 tons in 1904. Export to Europe was growing for certain types of apples, but exported apples went on the train to Montreal, not on the Murray Canal. The apples on the canal were destined for large evaporators in Trenton and Belleville, which produced dried apples, a very popular item in grocery stores.[13]

After seventeen years of operations, the entrance piers at each end of the canal would be replaced in 1906. The Ministry of Railways and Canals reported on September 14, 1906, that "tenders [had] been sought for concrete superstructures for the wooden piers at the entrance of the Murray Canal, and reinforced concrete bridges to span the openings between the piers."[14]

There were thirteen responses, the estimates ranging from $54,365 to $81,156. The lowest bid had been submitted by Robert Weddell, for his Trenton Bridge and Engine Company, and it was accepted.

There was some confusion with the first tender, dated June 20, 1906, due to an incomplete description of the work to be done, so a second tender was issued in September, with Robert Weddell presenting the lowest bid.[15]

This was Robert Weddell Jr., who had taken over the business after his father had passed away in 1898. There was a good start on the work before winter set in, but it was obvious that the work would take longer than expected, so the Ministry of Railways and Canals extended the contract to June 1, 1908.[16] The work on the piers was completed in November 1908.[17]

Robert Weddell Jr.

The Central Ontario Railway took another crack at supporting traffic to Twelve O'Clock Point in 1908 when they asked the Department of Railways and Canals to lease two acres of land along the north side of the canal so that they could build a spur line from their tracks farther west. The electric railway discussed a few years early came to nothing, but there were still big crowds of people coming to Twelve O'Clock Point by land in their wagons and buggies. While this second request was approved in a memorandum dated Sep 16, 1908, it does not appear as if it bore fruit.[18]

During the first decade of the 1900s, complaints grew about the depth of the canal and its approaches. According to the Powles Report, "throughout the winter and spring of 1909, the government was besieged by complaints and petitions requesting that the canal be deepened to 14 feet. These came not only from the cement companies, but also from steamship lines like the Mutual Steamship Company. They were very active in the Bay of Quinte area, and reportedly shipped over 20,000 tons of freight out of Belleville in 1908."[19]

E. G. Porter, a local member of Parliament, harkened back to the hope for the canal, claiming, "[T]he Bay of Quinte is part of the most important waterways in the Dominion, and has a very large carrying trade, which, owing to the establishment of certain industries along its waters, is constantly increasing and will increase more rapidly in the future."[20]

Due to the growing public pressure, surveys of the canal were carried out by the Department of Public Works in the early spring of 1909. "While these found some points in the canal which had filled in to about 9 1/2 to 11 feet of depth, this was at the low water level of Lake Ontario. At normal water levels, the waterway offered a clear navigation depth of 13 to 14 1/2 feet." This report did not identify a serious problem so nothing further was done.[21]

At the same time, the Belleville and Lehigh plants were taken over by Canada Cement, a new company "formed by a combination of the financial interests which owned ten large mills in Canada."[22] Extensive upgrades were done at both Point Anne plants with a view to increasing production. In light of this, pressure increased on the government to address the Murray Canal, which was perceived as an impediment to the traffic anticipated in the next decade. Belleville business interests became involved during 1910 by actively petitioning the government to take action.[23]

An incident occurred in Presqu'ile Bay, which the lobbyists used to press their case. "There were reports of a freighter which had run aground at Salt Point Light and in working itself free had spilled gravel into the dredged channel. This left a shoal of barely nine feet depth, which together with other obstructions in the canal and approaches posed a serious impediment to their freight trade in this region."[24]

These problems were being felt more keenly by the shipping community because the volume of traffic on the canal had increased dramatically, largely due to the cement plants. In 1910, the total tonnage was close to 178,000.[25]

Further surveys were carried out in 1910 that "revealed substantial shoals and high places in the canal cut and approaches." In light of this, tenders opened in 1911 for the work required to clear the canal of these obstacles.[26] The Powles Report explains that "the contract was subsequently awarded to W. E. Phin of Welland, who began work in September. Throughout the fall the canal was dredged east of the Trenton Road bridge and in the east entrance, and boulders removed east of the Brighton Road bridge."[27]

A second contract in 1913 with Macdonald Contracting Company of Toronto dealt with the high spots and a few boulders at the west end of the canal and in Presqu'ile Bay. This work would alleviate the problem of vessels grounding in the canal and its approaches.[28]

Some of the surveys in this period addressed the potential of dredging the canal to fourteen feet, but the work was never done. Part of the reason for this was that, in 1914, Canada Cement closed the Belleville Portland plant and limited its operations to the Lehigh site, a step that reduced traffic in the canal.[29]

In fact, by the time World War I came around, the age of the canal had played itself out. The railway system had developed across the province to make rail traffic more efficient for many bulk items. Coupled with the growing size of vessels, the desire of large shipping companies to use the Murray Canal would diminish.

It was not uncommon, especially in busy times, for vessels to collide with bridge piers. However, a slightly different event took place on August 18, 1912, when a stone barge from the Point Anne quarries collided with the wharf near the Smithfield Road bridge. The barge sank in the canal, blocking traffic. The owners inspected the vessel and decided that salvage was not feasible, so it was blown up underwater, with the remains removed by dredging.[30]

The very next year, the canal experienced a major event during the very busy navigation season, which interrupted traffic for a time. "The most serious accident occurred on May 10, when the three masted schooner *Major N.H. Ferry* struck the railway bridge pier and sank just east of the Smithfield bridge. The raising of this vessel was quite difficult and it was necessary to close the canal for several days to allow the salvage crews to work. It was finally removed on June 10, and sunk by the owners in the Bay of Quinte about one mile east of the canal entrance."[31]

To the west end of the canal, in Presqu'ile Bay, the Salt Point Lighthouse was finally demolished in April of 1913. This light had been deemed redundant after the Brighton range lights had been installed in 1891 to guide ships to and from the Murray Canal. Local residents lobbied for the continuation of the old Salt Point Light, but when its natural end-of-life date was reached in 1913, it was demolished, signaling the end of an era on Presqu'ile Bay.[32]

Repairs were required to the railway bridge in 1915 after a train passed over the bridge before it was completely locked into place. Railway workers were able to repair the damage to the rollers in order to keep the bridge operating, although it was highlighted at the time that the bridge needed more serious work.[33]

In 1917, some very small but important changes were implemented that would be noticed by people driving over the canal. Reduced traffic on the canal was accompanied by increased traffic on the roads. As a result, navigation accidents were replaced by road accidents. In the early days of the canal, the only warning for drivers on the road was a red light on the bridge. "As road traffic grew, there were several instances when drivers failed to notice that the bridge was open and barely avoided skidding into the canal. In an attempt to eliminate this, chain barriers complete with a red light were installed at the bridge approaches in 1917."[34]

The reduction in canal traffic was evident in the numbers. In 1918, only 44,735 tons of freight passed through the canal. Within seven years, the tonnage was 46,703, but only 1,200 tons represented commercial freight. Larger vessel size and the railways contributed to this change, but the common use of automobiles by the general public added to the mix. By the 1920s, private automobiles became the primary means of leisure travel, eventually putting an end to the pleasure steamers, even the very popular *Brockville*, which had replaced the *Varuna* before World War I.[35]

A story about the *Brockville* circulated around the community in the early 1920s, according to the Powles Report:

> One night in the early 1920s, Ernest Green and his father, who lived near the Trenton Road bridge, awoke to the sound of a shrieking ship's whistle. After running down to the canal, they found that the steamer *Brockville* had stopped just short of the bridge, which remained closed. The bridgetender on duty was a World War One veteran by the name of Scotty Robertson. A well-liked fellow who apparently enjoyed an occasional drink, he was discovered unconscious on the floor of the watchhouse, undisturbed by the ship's whistle. As this was not the first time this had occurred, Green and his father were familiar with the operation of the bridge and they let the vessel through. They continued to operate it until Robertson could resume his duties.[36]

The railway bridge received some further attention in 1919 when a survey was conducted. The general idea was that the bridge was old and should be replaced. However, no action took place as a result of the survey. In hindsight, the survey may have been a routine administrative step by Canadian National Railway after they took over the Central Ontario from the unstable Ontario Northern Railway. In any case, the bridge remained as it was.[37]

Increased road traffic meant that the road bridges required more attention. In 1920 and 1921, all three road bridges were re-floored with new timber. Only four years later, the work was repeated for the Trenton and Brighton Road bridges,[38] which demonstrated that the materials and technology were not up to the task of supporting the constant traffic of horse-drawn vehicles as well as the newer wheeled automobiles and trucks. This problem would develop over the next decade as maintenance practices had to change to meet new conditions.

It was apparent that the government saw the Murray Canal as a local waterway, and this was demonstrated clearly when, in 1926, the canal was officially transferred from the St. Lawrence system to the Trent Canal. On both the Trent Canal and the Murray Canal, traffic would remain sparse for the next couple of decades, until recreational traffic began to increase in the 1950s.[39]

In 1927, the aging of the canal infrastructure became evident when the timber crib work between the pivot and rest piers showed signs of rot, which threatened to release stone ballast into the canal. This was a serious matter and the timber structures were "replaced by a 114' long concrete wall on the navigation side, while a dry wall was built on the other side of the pier to retain the stone filling."[40]

CHAPTER 13
Problems with Bridges

Smithfield Road bridge collapsed, August 30, 1930

A few years later, a serious event occurred at the Smithfield Road bridge. In August of 1930, the bridge collapsed. Repairs were underway to replace rotten timber bracing, and during the work, they had to open the bridge slightly. The south-west corner of the bridge collapsed under the stress. This picture shows the bridge soon after it had collapsed. We are looking north, to the bridgetender's house on the north side of the canal, and can see the collapsed south-west corner of the bridge. After some effort, they were able to swing the bridge to a fully open position so that traffic could continue on the canal and the authorities could take time to evaluate the situation.[1]

Eventually, in 1933, the decision was made to discontinue the Smithfield Road bridge. The price tag was too high, and the two remaining bridges could handle the traffic. The center piers were left intact, but the bridge was dismantled with all useable parts stored for future repairs to the other bridges. As a result, the old Smithfield Road was closed, marking the end of an era.[2]

Smithfield Road Bridge pier today at end of Hutchinson Road

This piece of road had been part of the original Danforth Road built from York to Kingston in 1800. The idea was to link the Smithfield settlement with the active commercial area of the Carrying Place, where Asa Weller had his trading post and batteau portage service. Two decades later, when the York Road was built as an upgrade to the Danforth Road, the new road was pushed directly east from Brighton to Trent Port, leaving the old Smithfield Road as a popular local route to the Carrying Place area. In the 1880s, a bridge over the Murray Canal allowed the road to continue in service, but in 1930, it was closed at the Murray Canal. Today, the gravel road down to the canal is called Hutchinson Road after a local settler family. You can see the remains of the piers of the Smithfield Road bridge in the canal. If you don't know the history, it looks very odd sitting there in the canal.

It was not only the road bridges that received attention in the early 1930s. An agreement had been reached by which the canal work force would be responsible for work on the railway bridge and the railway workers would deal with the tracks. In the early 1930s, many railway ties were replaced along with a significant amount of timber in the bridge superstructure. Similar to the work on the other bridges, the railway bridge saw major renovations of the timber crib work of the substructure, which were replaced by concrete retaining walls.[3]

A tragic event occurred in 1932, which caused a hard look at safety measures for road traffic approaching the canal. Through the 1920s, the switch of traffic volume from canal to road caused the bridges to be left open for road traffic. This, in turn, caused many drivers to develop the habit of maintaining speed as they approached the bridge. This was particularly dangerous at the Trenton Road bridge because the road made a sharp turn just before the southern approach to the bridge. Warning signs had been installed, but the potential for tragedy remained.[4]

On the night of November 22, 1932, conditions came together to cause a northbound car to skid into the canal. Three people were in the car but only one survived. This event led to the installation of much better warning signs as well as new guardrails up to the bridges.[5]

Bridge repairs continued in 1935 when the Brighton Road bridge had the original suspension cables replaced. The next year, extra bracing was added to the bridge trusses. Concrete work was also done at the base of the retaining walls of the piers.[6]

The year 1936 saw the final realization that the road bridges must be much more structurally sound in order to withstand the constant pounding of modern road traffic. The Trenton Bridge required a lot of repairs, and the decision was made to replace the bridge. There were multiple considerations that went into the design of the new bridge. Certainly, they needed to build a bridge that would remain structurally sound under the immense pressure of vehicle traffic as it was developing in the 1930s. Besides that, it was thought necessary to have a bridge that provided two full-sized lanes of traffic as well as a pedestrian walkway.[7]

The Ontario Department of Highways carried out the work while the federal Department of Railways and Canals contributed $24,000 to the project. In fact, this was a combination road and canal project since the Trenton Road had to be straightened in conjunction with the new bridge being built. The location of the bridge would be a few hundred yards east of the old bridge, with much smoother access for the road at both ends of the bridge. The work began in the fall of 1935 after close of navigation. Cameron Phinn of Welland had the contract for the substructure and abutments, and the steel superstructure was built by Mr. Hill, a local contractor who was later involved in the Central Bridge Company of Trenton.[8]

Extreme cold over the winter allowed Phinn's men to use the ice of the canal as a solid floor, which moved the work along quickly. The Powles Report notes that "by early spring they had completed the rest piers, which consisted of stone filled cribs built to the water line, then topped with concrete monoliths. The south abutment, along with a portion of the pivot pier was also completed, to be connected to the rest piers by pile cribs."[9]

The work was finished in the summer of 1936, and the bridge was opened for public use in November. "It consisted of a 146' long Through Steel Truss span, with a reinforced concrete approach-way on the south bank. In contrast to its one lane predecessor, this bridge boasted a 20' clear roadway, as well as a sidewalk which was bracketed to the bridge truss."[10]

Bridge carrying Highway 33 over the Murray Canal

Those who were more old school about civil engineering projects may have felt that this bridge was far too expensive and elaborate for a simple country road crossing. However, the presence of two full traffic lanes and a safe pedestrian walkway was becoming the standard. As one local resident put it, "sometimes you were taking your life in your hands to walk across the old one, especially when some of the hotrodders went around."[11]

Now, the bridge that took the Carrying Place Road across the Murray Canal was much safer for both vehicles and pedestrians. Note the name change. The name of the post office was changed from "Murray," the traditional name which reflected the township, to "The Carrying Place," which reflected the historical reality of the isthmus. The change came about after local citizens lobbied the post office, including a very convincing letter from William Hodgins Biggar, who had recently moved to Montreal for his job as general counsel for the Grand Trunk Railway Company of Canada. He was also a son of James Lyons Biggar, the staunch advocate for the Murray Canal.

Mr. Biggar presented the request to the postmaster general in Ottawa, in a letter in 1913, saying, "I have just received a letter from the Postmaster of Murray Ont., stating that a petition is before you asking that the name of this post office be changed to 'The Carrying Place.' As I was born at 'The Carrying Place' and both my Grandfather and Father for years filled the position as Postmaster, I naturally have considerable interest in the matter." Again, lobbying and advocacy in the right way from the right people had an effect and the name was changed.[12]

As to the bridge itself, there was one unintended consequence of replacing the old composite steel and timber bridge with a new steel bridge. The bridgetenders soon realized that this new bridge was a lot heavier than the old one. In fact, it became a problem for the bridgetenders because they were having a lot of difficulty turning the big handles that swung the bridge. With the old bridge, it had required little effort, but now, it was a chore.

Several of the bridgetenders were retired war veterans, so this added physical burden was not easy to deal with. In some cases, they would hire young men from the community to push the big handle to turn the bridge. It would take a couple of teenagers to do the job, but they were willing to do it for ten cents a swing or fifty cents for all the swings in a day. In the early 1940s, another approach was tried. A tractor was brought in, connected to the big handle with a rope, to open and close the bridge.[13]

The Brighton Road bridge would finally receive some attention in 1946. It was apparent that both substructure and superstructure of the bridge needed major work. However, the decision was made to replace the bridge in order to increase the five-ton capacity limit, which was considered inadequate to support existing traffic.[14]

Work on the substructure began early in 1946 with the hope that it would be done before navigation season opened in the spring. The Powles Report states that "canal forces began to remove the old masonry and crib work piers to the waterline. They were then rebuilt in reinforced concrete and enlarged somewhat to accommodate the new bridge." Unfortunately, the work delayed the opening of the canal until June 23 and was not completed until July 19.

The superstructure was built by the Central Bridge Company of Trenton. The bridge was erected in the open position beginning in August 1946 and opened for traffic on November 13, 1947. The total cost of the work was $59,955, with $18,220 of that total representing the involvement of the canal work force in the replacement of the substructure.[15]

New bridge at Brighton Road in 1947

The new Brighton Road bridge was almost identical to the Trenton Road bridge, including its weight. The bridges were still swung manually, and the heavier bridge at the Brighton Road not only resulted in headaches for the bridgetenders, but also caused delays for road traffic trying to get across the canal.

The obvious solution to this problem had developed over time in the wider world as electricity became a common tool to assist mechanical systems. It then became a matter of deciding at what point the problem began to impact the efficiency of the system. Only then could budgetary considerations be accommodated. "Not surprisingly, in 1947 it was decided to electrify both bridges and the following year the necessary equipment was purchased."[16]

Over the next several years, the implementation of electrical systems to turn the swing bridges would run into delay after delay. Late delivery of the new equipment caused a postponement from 1948 to 1949. Some equipment was installed in 1949, but the work was interrupted again because of a dispute over whether to use mechanical or hydraulic drive units. The decision was finally taken to use mechanical units, and installation of the equipment along with the control panel was completed during 1950,[17] to be fully operation in 1951.[18]

At the same time, new traffic gates were installed with a view to driver safety. When the new systems were ready for operation in 1951, it was found that drivers coming up to the bridge might confuse the lights on the traffic gates with the navigation lights on the bridge. The solution to this problem was a shield placed over the navigation lights that helped drivers focus on the traffic gate lights.[19]

In spite of the all the early problems, this new method of swinging the two road bridges worked very well. The bridgetenders were happy because one difficult part of their work was eliminated. Drivers on the roads were happy because the bridge swinging procedure was quick and reliable. Pedestrians were also happy because there was now a safe walkway on both road bridges.

The electrification of the bridge mechanism began a period of uninterrupted service and minimal maintenance for the Murray Canal. In 1958, electric navigation lights were installed on all the bridge center piers in order to facilitate navigation at night.[20]

Then, in 1967, the drive motors that powered the bridge-swinging mechanism burned out at both bridges. "During the off-season, both of these units were sent to the main shops at Peterborough to be modified for a change over to a new hydraulic operating system. These were installed over the season of 1968–69, and were fully operational by the start of the 1969 navigation season."[21]

But then, a serious interruption to traffic occurred at the Brighton Road bridge on June 7, 1969, during the busiest part of the season. A malfunction of the new system caused a delay of nine and a half hours. Lessons were learned at every step of the process, and new methods and approaches were implemented as a result. During the later part of the 1969 season and continuing into 1970, controls for the bridges were placed inside new control booths, which would act as the office of the bridgetender.[22]

The next major work on the canal was at the Carrying Place bridge during the fall and winter of 1977 and 1978. Baltimore Development Services of Cobourg removed the decaying fixed concrete span along with sections of the abutments and center pier and replaced them with new reinforced concrete structures. The project cost $49,000 and, with delays, was not finished until well into the winter of 1978.[23]

The emphasis had been on the road bridges for several decades while the railway bridge continued to function adequately. However, this would change, as "in the early 1950s, heavier traffic began to utilise this line, particularly iron ore shipments from Marmora to Picton. Apparently, this placed a great strain on the existing lattice bridge, so in 1954–55 it was replaced by a heavier plate girder structure. It seems that this was not a new bridge, but was moved from the Third Welland Canal and installed by CN railway forces."[24]

The railway bridge was in place by the spring of 1955, but there were many delays related to the electric bridge swinging system. The first system installed was simply not powerful enough to swing this new and much heavier bridge. As a result, the bridge had to be swung manually for most of the 1955 season. This required two bridgetenders and two canal employees, who were temporarily assigned to assist with this problem.[25]

The Powles Report provides the following from a local man named Allan Alyea, who worked in this situation: "It took four men to swing the damn thing. It was just like walking to Brighton when you started to open it."[26] This problem was resolved in 1957 when a larger drive unit was installed and a limit switch was implemented to regulate the movement of the bridge during operation.[27]

The next improvement at the railway bridge was in 1966 when "an auxiliary set of controls was placed on the south bank of the canal to allow the bridge to be swung from the shore."[28] This mirrored the approach that was being taken at the road bridges. Complex electrical systems needed control panels that must be housed in protected control booths. These booths became a common sight at all the bridges over the Murray Canal.

The railway bridge saw the replacement of a drive shaft in 1968 and renovations of the drive system in 1972.[29] However, during the 1970s and 1980s, traffic across the bridge declined, and at the writing of the Powles Report in 1991, it was only used twice a week "to allow the passage of several cement cars from the Lake Ontario cement plant in Picton."[30]

There is a picture of the railway swing bridge over the Murray Canal on page 90 of the book *Desperate Venture*. The picture was taken by the author of the book, James Plomer, in 1977. It shows the railway bridge swung open in its perpetual position, still showing rails approaching the canal from the south. The bridgetender's control booth is still there, and we can even see the X of the railway sign on the road, just behind the booth.

CHAPTER 14
People of the Canal

A history of the Murray Canal must include information about the people who lived along the route of the canal. "Land Expropriation" gives details of land expropriation for the people along the route. "Workers and Their Homes" provides stories of some of the people who worked on the canal from maintenance workers to bridgetenders.

Land Expropriation

The people most impacted by the construction of the Murray Canal were those who owned land along the route. There are forty-three transactions related to the expropriation of land for the Murray Canal in the Ontario Land Registry Records. Appendix F shows all the land transactions, and there are Orders in Council in the Library and Archives of Canada to match most of these transactions.

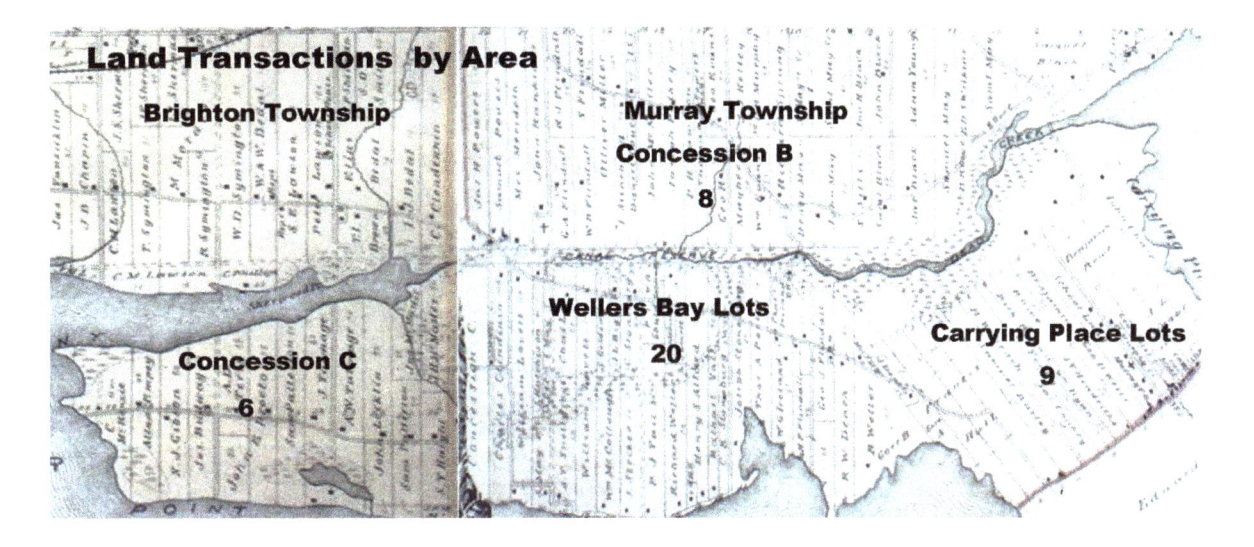

Land transactions by area

The transactions break down by location with twenty from the Wellers Bay lots south of the canal and toward the west end, nine from the Carrying Place lots at the east end, eight from Murray Township, Concession B, toward the north side of the canal, and six from Concession C, Brighton Township, which is better known as Stoney Point, at the far west end of the canal.

The Carrying Place lots had been surveyed when the first settlers established themselves along the narrow isthmus that linked the Bay of Quinte and Wellers Bay. Asa Weller was there first, so the lots were numbered from his trading post at the west end of the portage. These lots were long and narrow, allowing owners a front on the Carrying Place.

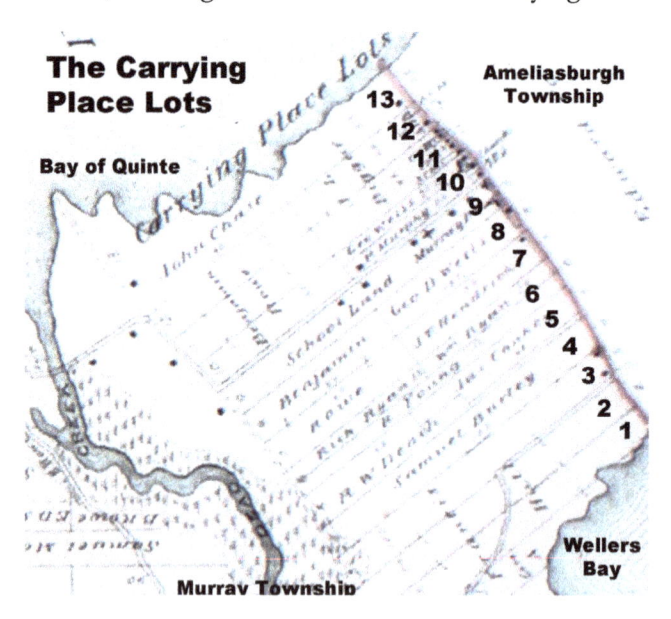

The Carrying Place lots

Starting at the east end of the canal route, Carrying Place Lot 13 is mentioned first, being the land right along the shore of the Bay of Quinte. Louis Latour had obtained part of the south half of Lot 13 from the Biggar family in May of 1883, then, in June, had to give up "2 776/1000 acres" for the Murray Canal for which he received $175.[1]

Latour had been a resident of Ameliasburgh Township and was married to Catherine Maria Bonter, whose family were among the earliest settlers along the south shore of the Bay of Quinte, where the road was called "The Bonter Sidewalk." He had acquired land in Carrying Place Lots 12 and 13 in 1878, accompanied by a good deal of wheeling and dealing with neighbors in the area.[2]

A much smaller piece, shown as "29/100 acre" was given up by David John Huffman, from his property in the north-west part of Lot 13.[3] This land was called a shoreline lot and had belonged to John Chase only a few years before. There was a lot of buying and selling of land in this area in the 1870s, which makes one wonder if folks were anticipating the route of the canal and making their bets.

The next three transactions, for Carrying Place Lots 12, 11, and 10, show that Benjamin Rowe and wife were paid $2,500 for land along the canal route. This goes west as far as the Trenton Road, which was on the lot line between Lots 9 and 10, going straight into the village of Carrying Place.[4]

Rowe had acquired land in Lot 12 in 1868,[5] and we don't have to look far in the records of many other Carrying Place lots to see that he was very active in this area in the 1860s and 1870s. He is listed as a farmer in the census records, but land speculation was a common activity as well.

Benjamin Rowe was born in Sidney Township in 1812, son of John Row and his second wife, Mary Meyers, who was a daughter of John Walden Meyers, the famous early merchant and trader at the town of Belleville. Obviously, he had deep and close ties around the Bay of Quinte.[6]

Carrying Place Lot 9 is clearly marked on the Belden County Atlas map of 1878 as "School Land," and since the canal route passed through this lot as well, the trustees of School Section 1, Murray Township, were the grantors of a small bit of land shown as "216/1000 acres."[7]

Benjamin Rowe and wife are mentioned again for Lots 8[8] and 7,[9] with only small slivers of land given up from the north part of the lots. From Lot 6, Thomas A. Porter gave up 92/100 acres from the north part of his farm.[10]

Looking a little closer at Thomas A. Porter, we see that he was married to Mary Jane Rowe, a daughter of Benjamin Rowe, and the records show that after his expropriation in 1884, the next year, he sold part of Lot 6 to his father-in-law.[11]

Continuing down the list leads us to Murray township, Concession C, which is the land north of Wellers Bay and south of the Canal Reserve. Many people had to give up land for the canal at the far north end of these lots.

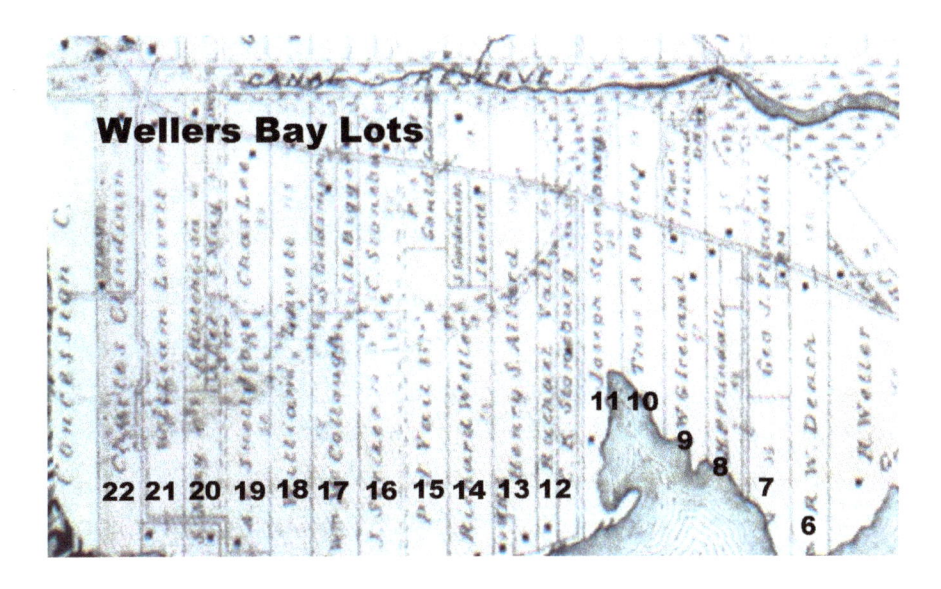

Murray Township, Concession C

Stephen H. Flindall and his wife gave up two acres plus 757/1000 acre from both Lots 8[12] and 9[13] and received $120 in compensation. The land records use the name Stephen H. Flindall, although he was commonly known as Henry. His grandfather, John Morris Flindall, had brought his growing family to Canada after the War of 1812 and obtained the Patent from the Crown for all of Lot 8 of Concession C, Murray Township.[14] By the 1870s, much of this land was still in the family, including with Henry's brother, George James Flindall, who gave up "134/1000 acres" from Lot 8.[15] This may seem to be a trifling bit of land to lose, but when it is considered part of your family's heritage, any amount in compensation would be seen as the trifle.

Transactions for Lots 10[16] and 11[17] were "decrees" from the Court of Chancery in favor of "Her Majesty the Queen." Both lots were in the hands of the Stoneburgh family in the 1860s. Lot 10 was sold to Benjamin Rowe in 1871, and the next transaction was the decree. Joseph Stoneburgh had granted property in lot 11 to his wife, Eliza Jane, per his will, and then the decree. In any case, the queen got her land for the canal.

Joseph Stoneburgh's grandfather, Peter Stoneburgh, had been born in New York State but came to Canada as a United Empire Loyalist, obtaining the Patent from the Crown for Lot 11, Concession C, Murray Township.[18] As was common in families during the early and mid-1800s, this land passed down generations from father to son. Peter passed it to William[19] and he passed it on to his son Joseph.[20]

Lot 12 had no land appropriated, and Lot 13 had a transaction for "west half 200th of an acre," given up by Peter and Hannah Gould.[21] This is where Dead Creek Marsh extends south and where the dredging was said to be in "Gould Clearing"[22] — just a couple of lots west of where the Smithfield Road crossed the marsh.

Hannah Maria Allard had married Peter Gould in 1845, both families being well established on farms on Concession C of Murray Township. Peter's grandfather, John Gould, had been a member of Butler's Rangers when that group was disbanded after the war in the Niagara area. John obtained land in Glengarry County, and his family spread out from there. Two of his brothers, Nathan Gould and Seth Burr Gould, ended up in Cramahe Township, near the present village of Salem.[23]

Hannah Maria's father, Enoch Allard, had come from New Hampshire to settled in the south end of Murray Township before 1820, to become neighbors of the Gould family. Henry S. Allard, Hannah's brother, had married Katie Lawson, a member of the early-settler family in what was called "Lawson Settlement," south of Smithfield.[24] He gave up his interest in a small amount of land in Lot 13 as well.[25]

Peter and Hannah Gould also gave up one full acre plus 62/100 acre in Lot 14, land that they had acquired in 1853.[26] A larger chunk, three and 13/100 acres of lot 15, was given up by William H. Goldsmith and his wife, who happened to be Sarah Gould, a daughter of Peter and Hannah Gould.[27]

As we progress west, closer to the Brighton Road crossing, the amount of land expropriated increases. Cadwell Ketchum Stoneburgh gave up four and 82/10 acres from his land in Lot 16.[28] He was a son of Abraham Stoneburgh and his second wife, Lydia Gould, a sister of Joseph Gould, Peter's father.[29]

Family connections among early settlers can be complicated. When dealing with family trees in the early 1800s, we need to keep in mind that people did not travel much and often married their neighbors. For those of us who love genealogy and history, this kind of analysis links those two disciplines together with real people in places we know. This is what makes it fun!

There were several transactions for Lot 17. Charles and Hester Lee gave up three and 32/100 acres for $125.[30] Hester A. Williams had married Charles Lee in 1876 and acquired twenty-five acres at the north end of Lot 17 in 1881.[31] One of her siblings was Emily Jane Williams, the wife of Cadwell K. Stoneburgh.[32]

Lot 17 had been divided at the north end some time ago, and another expropriation saw three and 32/100 acres given up by Mary Goldsmith and husband.[33] This was Mary McKenzie,[34] wife of Gilbert Goldsmith, and these two were the parents of William Henry Goldsmith mentioned regarding Lot 15. Another two and 95/100 acres was given up from Lot 17 by Phillip H. Lawson.[35]

Approaching the Brighton Road, William Lovett owned all 130 acres of Lot 18, which he had acquired in 1851.[36] Now, he was forced to give up a sizable chunk of the north end of this property, described as "7 and 47/100 acres" for which he received $500.[37]

Charles and Hester Lee were neighbors, on Lot 19, and they gave up slightly more, "7 and 63/100 acres" with compensation of $600.[38] The Order in Council documents that match these transactions often included details that can help explain the compensation. Sometimes, they took into account how much swamp or arable land was being expropriated and how much woodlot was on the land the farmer was giving up.

Next, we see that Samuel May and his wife gave up a small amount, 24/100 acre of their property in Lot 20, receiving $20.[39] Also from Lot 20, Jonathan Hutchinson gave up a large piece of land, six and 13/100 acres, for $3,000.[40]

Joseph Wilson and wife gave up one and 99/100 acres from the same lot for $150.[41] His father, John Wilson, had been a very early settler on Concession C of Murray, obtaining the Patent from the Crown for Lot 16. They were close neighbors to the Gould family, and Joseph Wilson had married Mary Jane Gould, a daughter of Joseph Gould and Rachel Stoneburgh.[42]

One of the larger expropriations came from Lot 21, another part of William Lovett's farm at the Brighton Road, where he gave up eight and 54/100 acres with compensation of $2,500.[43] The last transaction in Concession C of Murray Township was in Lot 22, where Charles Clindinin and wife gave up seven and 58/100 acres for $1,000.[44]

Charles Clindinin had obtained this property from his father, James Nelson Clindinin, in 1866,[45] and this expropriation would be one of several dated December 29, 1882, the earliest batch of transactions for the Murray Canal. The rest would be dated from January to May of 1883.

The next group of transactions deals with Concession B of Murray Township, the land immediately north of the Canal Reserve. It can also be seen as the northern mirror of Concession C, as the lot numbers line up across the canal.

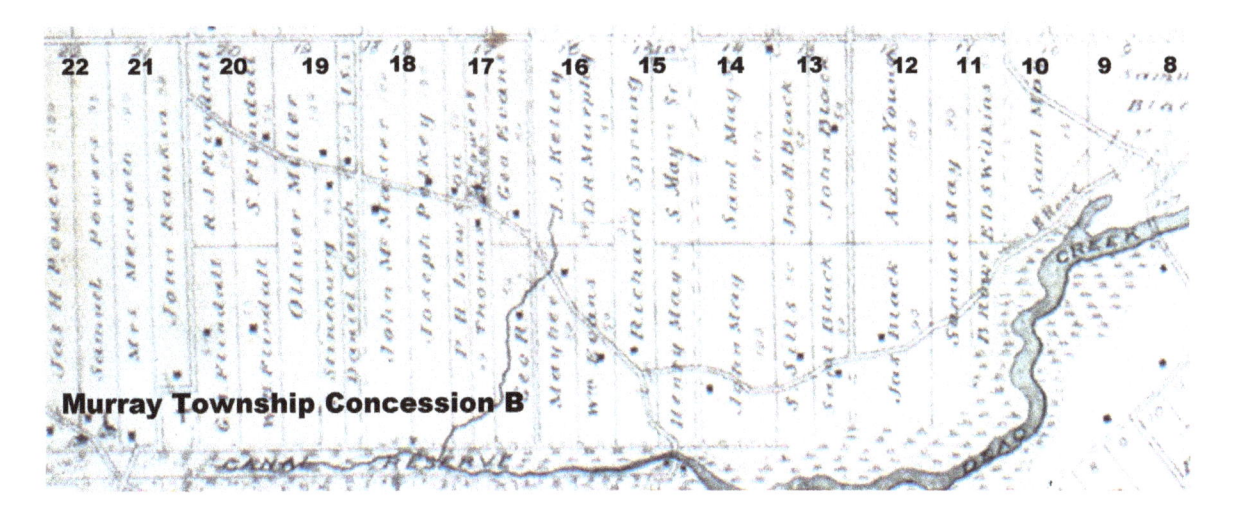

Murray Township Concession B

Only Lots 13 to 18 of Concession B are included in the list of land transactions for the Murray Canal. Less than half an acre at the south end of Lot 13, shown as "545/1000 acre" was given up by Sylvester Sills with compensation of $40.[46]

Samuel May and wife gave up a bit more from Lot 14, "844/1000 acre" for $385.[47] It should be noted that this was Samuel May and his wife, Lucy Hetherington,[48] a different Samuel May than the one mentioned for Lot 20, Concession C.[49] Following this theme, a son of Samuel and Lucy May, John May, gave up a large chunk, five and 844/1000 acres, in Lot 14 of Concession B.[50]

Three transactions were made for Lot 15. Looking at the map, we can see that Lot 15 contains the place where the Smithfield Road came down to the Dead Creek Marsh. A bridge had carried the road over the creek at this spot for decades, and a new bridge over the Murray Canal, called the Smithfield Road bridge, would continue that job.

Another son of Samuel and Lucy May, George H. May, gave up two acres and 65/1000 acres from Concession B, Lot 15 for $180.[51] Two more transactions on the list refer to Lot 15, but they are called "Certificate, Vesting Order." One is for the same two and 65/1000 acres given up by George H. May mentioned above.[52] The other is for a larger piece of land, six and 838/1000 acres.[53] No dollar amounts are shown for these Vesting Orders, but it would appear that the queen received the land she needed, whether there was legal wrangling or not.

Samuel Fletcher May was an unmarried man when he gave up a substantial piece of his farm in Lot 16, described as "4 98/100 acres" with compensation of $210.[54] He had acquired fifty acres in the south part of Lot 16 only a year before, from his parents, John May and Frances Powers.[55] Here is another place in the land records where it seems apparent there is a bit of jockeying for position going on in anticipation of the canal coming this way.

On the other hand, a very small bit, "84/100 acres," would be given up from Lot 17 by Thomas Potts Powers and wife, along with this widowed mother, Elizabeth (Potts) Powers.[56]

The last expropriation for Concession B of Murray Township was in Lot 18, for only 0.54 acres, written as a decimal number, which is an odd exception, considering all the other items. The grantor was Joseph Thomas Pelkey and his wife, Catharine Stoneburgh, a daughter of William and Clarissa Stoneburgh.[57]

One last area experienced significant expropriation of land at the west end of the canal route. These transactions are for Concession C, Brighton Township, or Stoney Point. This is the peninsula that juts out into Presqu'ile Bay across from the lighthouse and forms the eastern side of the entrance to the bay. Along the north side of Stoney Point is Weese's Creek, the eastern extension of Presqu'ile Bay. Only Lots 23 to 28 are affected here, representing the western entrance to the canal. This map segment shows Stoney Point oriented east and west, north is to the right and south is to the left.

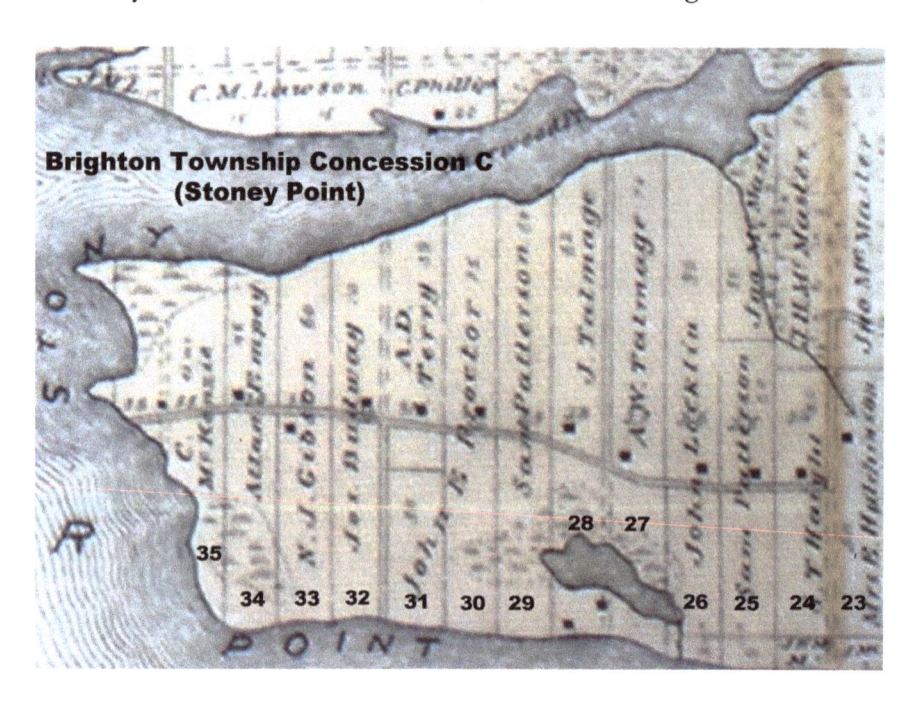

Brighton Township, Concession C

John McMaster and his wife owned the north half of Lots 23, 24, and 25, right where the canal would enter Presqu'ile Bay. Here is the text in the Order in Council:

> On a Memorandum dated 9th October 1882 from the Acting Minister of Railways and Canals recommending upon the report of the Government Land Valuators, that authority be given for payment to Mr. John McMaster Junior on proof of title, of the sum of one thousand dollars ($1,000) as compensation in full for parts of lots 23, 24 and 25, concession C, township of Brighton, consisting of about 18 acres, expropriated for the purpose of the Murray Canal, and for loss by depreciation in value of a further quantity of land 15 acres in extent — and that the said amount be charged against the appropriation for the Murray Canal.[58]

Just to be clear, this was John McMaster (1818–1888) and his wife, Mary M. Manley (1820–1906). He was born in Sidney Township and was a brother of Samuel and George McMasters, although his records tend to show his surname spelled without the "s" on the end. Anything to confuse researchers![59]

John McMaster had acquired the north part of these three lots starting in 1847 when he purchased two acres in Lot 22, where the Brighton Road came through.[60] This may have been seen as a strategic location for a store or blacksmith shop.

Then, in 1852, he purchased fifty acres of the north part of Lot 23 from Joseph A. Keeler of Cobourg, who was the father of Joseph Keeler III, the advocate for the Murray Canal.[61] This land included the south side of the end of Weese's Creek and was beside his previous acquisition in Lot 22. The process continued in about a year when John McMaster purchased fifty acres in the north half of Lot 24,[62] and 45 acres from the north end of Lot 25,[63] extending his land further up Weese's Creek to the west.

Much of this land, especially toward the north side, was swampy, but John McMaster would have enough arable land to make a decent farm, while providing resources for fishing and hunting that might generate some income. Then, when decisions were finally reached to build the Murray Canal into Presqu'ile Bay, he was well positioned to collect some return on his investment.

Expropriations from lots west from the McMaster property were less than an acre each. Hiram G. Lawson and his wife gave up "97/100 acre" in Lot 26,[64] Aaron W. Talmage and wife gave up "644/1000 acre" in Lot 27,[65] and Daniel and Martha Church gave up 0.556 acres in Lot 28.[66]

Enough with the land expropriations!

Workers and Their Homes

Everyone in the immediate vicinity of the new canal had to adjust to the big ditch. Farms gained new fencing to define the boundaries along the canal for both humans and animals. Travel habits developed to utilize the three road bridges, no doubt with cursing from the older folks who simply did not want to see things change. Kids within a mile or so of the canal would soon find a new place to swim and fish, to say nothing of the entertainment value of all the sailboats, steamer ships, and barges that used the canal from the first day it was open.

The book *Gunshot and Gleanings of the Historic Carrying Place, Bay of Quinte* provides the following memories of people who lived and worked on the canal:

> Mrs. Clark recalls the names of some of the bridge masters. Ralph Jones is the one she remembers when she was a child, around 1908; then Will Johnson, Mr. McColl, Jack Scriver, John Hanna, Ben Young, Russell Simpson, William Horsely. The superintendents she remembers were Bert Richards, Joe Dickson, Col. Harry Sauve, Hugh Maitland and Charlie Brown. Bridge tenders who come to her mind were Ernie Barker, Mike Paro, Herb Hanna, Jim Montgomery and Jack Harvey. Thelma Schriver Alexander remembers other bridge tenders, men who worked on the middle bridge which was later removed. They were Frank Hodge and S. McQuinn. The superintendent who came after Bert Richards was Carmen Baker. The Scrivers lived for eight or nine years near the middle bridge in the second house on the English Settlement Road when Mrs. Alexander was a child.[67]

These names span several decades of canal operations, but let's look a bit more closely at some of these and a few others.

John Hanna grew up near Orland, a son of Richard Hanna and Martha Davis.[68] His uncles, George and John Hanna, had farms across the border in Murray Township. John Hanna and his wife, Susan McPhail, had a family of nine children in 1915, when he managed to land a job on the Murray Canal as a bridgemaster. They all moved to Carrying Place, where we can see them in the 1921 Census. They are in good company, since John Scriver, a bridgemaster, and Sam Stanton, a bridgetender, are just above. For good measure, John's oldest son, Hugh Burton, was a lock tender on the Trent Canal.[69]

A cousin of John's, Harold Grant Hanna,[70] was a World War I veteran who had been born at Grafton in 1898 but was part of the large Hanna clan stemming from his grandfather Richard

Hanna, who had settled near Orland in the 1820s. Harold lived in Smithfield after being married in 1934 and then he obtained the position of bridgemaster on the Murray Canal. He would be replaced in 1952 by Arnold Steenburgh.[71]

William Thomas Horsley was born in the Wooler area in 1898 and was known in the community as "Willie." He would be a bridgemaster on the Murray Canal for almost two decades, from early in the 1930s to the early 1950s.[72]

John Robert Harvey was born in Brighton Township in 1917, a son of Gordon John Harvey and Ethel Chatten.[73] In 1943, he married Beatrice Ellen Terry, and by 1958, the family consisted of five growing and active boys. While life on the canal would have been ideal for the boys, it certainly posed a challenge to the parents who chose to live in a house provided for the workforce. John Harvey's salary as a bridgemaster was welcome, but it did not ease the tight conditions in the house.[74]

Allan Alyea[75] is mentioned earlier in these pages as someone who worked on the canal. In fact, he lived close by so he would not need to use any of the tenant housing provided along the canal. His grandfather James Henry Alyea purchased the north half of both Lots 7[76] and 8[77] in Concession C, south of the Canal Reserve in 1885 from George J. Flindall, while the Murray Canal was being built. His father, Fred Alyea, farmed there all his life, and Allan continued the tradition. This property was located south and east of the canal at the Smithfield Road.

Lovett Canal House, c. 1900

A unique image comes from the Lovett family history showing a gentleman sitting in front of the door of what is labeled as the "Lovett Bridge Canal House." The hill immediately behind the building confirms that this is the Brighton Road crossing and appears to be one of the small buildings they built for the bridgetenders. For a time around 1900, Allen Edwin Lovett, also called Marcus, was a bridgetender for the Brighton Road bridge. It is thought that he is the fellow in front of the open door in the image.[78]

Marcus was a farmer right there south of the Brighton Bridge, which made it very handy for him to moonlight at the canal. He was a son of William Lovett, who had lost six and 13/100 acres of land at the north end of his farm when the land was expropriated in 1882 for the building of the Murray Canal.[79]

Soon after the canal was opened, Superintendent Keeler expressed the need for accommodation for the bridgetenders near the canal. He identified two houses close to the canal that would serve this purpose, starting with one house that had been built in 1891 by Wesley Goodrich. The Department of Railways and Canals initially leased these houses but would purchase them so they could be available as required for canal staff.[80]

Accommodation for the bridgetenders was an on-going problem. In 1902, the farmhouse built earlier by Wesley Goodrich was put up for sale, and the minister of Railways and Canals said:

> It is considered desirable to provide house accommodations for the bridge tender of the Smithfield Bridge over the Murray Canal, and for this purpose to acquire a residence with outbuildings including a barn, built in 1891 for Mr. W. Goodrich, on the Canal dump at that point, and which he offers to sell. The sum of $500 is suggested by the Superintendent of Operations as the price to be paid for the whole property. The Minister recommends that authority be given for purchase accordingly.[81]

This house was used by the Smithfield bridgetender until about 1912 when the canal labor foreman moved in.[82]

Another house was dealt with in 1904, when the minister said:

> In the year 1895, Mr. W. H. Johnson, Bridge Tender on the Murray Canal, built for himself a farm house on the Canal Reserve near the railway bridge, houses in the vicinity of the Canal being scarce, and no village being situated within 5 miles of it. The Minister further represents that subsequently he left the service of the Government, and having no further use for the house, has offered to sell it to the Government. He claims that the actual cost of the house itself to him was $1,200.[83]

In the end, the house was purchased for $850.[84]

The year 1908 saw the erection of a third dwelling north-west of the Smithfield Road bridge. It was a one-and-half-storey frame building with a kitchen, six rooms in total. Much of the material had come from the old toll collector's building that had recently been demolished. This house was used for the Smithfield bridgetender.[85]

The Department of Railways and Canals would take little responsibility for these buildings after they were built. Interior maintenance or renovations were up to the tenant, although the department would pay for materials. In effect, the department acted like a traditional landlord, which meant the buildings were not well maintained. The arrival of new tenants might motivate some fixing up, but otherwise, it was thought that maintenance was only needed every seven years. As a result, the tenants were very much on their own.[86]

According to the Powles Report, "living conditions in the other homes remained relatively spartan until well after the Second World War, when extensive work was done on the rapidly aging buildings. During 1948 and 1945 dwellings #52 and #53 were fitted with electric lighting, but only after considerable pressure from their tenants, H.G. Hanna and W.T. Horsley. Horsley also managed to get his house wired for electricity the following year, with eight percent of the value of the work added to his rent to cover the costs."[87]

Then "a similar arrangement was also made at dwelling #54 during this period. Attempts were also made to improve the heating of these buildings, and in the early 1950's [sic] coal furnaces were placed in houses #51 and #53. Indoor plumbing came next in 1954 and 1955 when both Smithfield houses and dwelling #51 were fitted with three piece bathrooms and hot water heaters for a cost of about $1000 each."[88]

By the 1960s, the policy changed regarding accommodations for canal staff. The buildings were aging badly, and the cost and complexity of providing on-site dwellings had been demonstrated over time. Besides, fewer workers requested local homes due to easier transportation and communication facilities. Everybody had a car and a telephone. As a result, the houses along the canal where bridgetenders and their families had lived for decades were eventually removed.[89]

CHAPTER 15
The Murray Canal Today

From the first thought of Simcoe's Canal in the 1790s, the canal was expected to be located in Murray Township, which was part of Northumberland County. Murray Township also included the village of Trent Port, later called Trenton, but that was changed in 1853 when the village of Trenton was incorporated and became part of Hastings County.

Then, in 1881, Trenton became incorporated as a town and became what was called a "separate town" within the county.[1]

The canal was built during the 1880s, entirely inside Murray Township. More than a century later, in 1998, a major change occurred that would impact Murray Township and many other communities in the area.

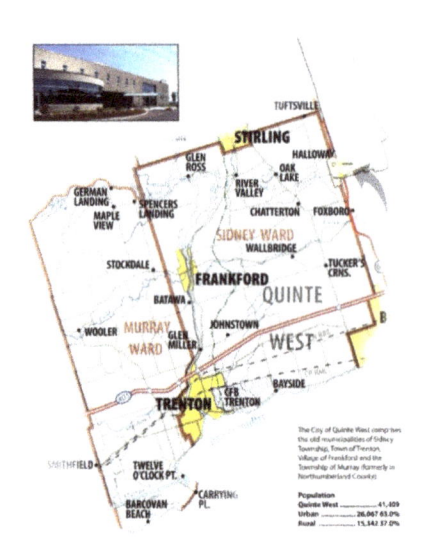

On January 1, 1998, Murray Township and the town of Trenton, along with other entities, were amalgamated into the City of Quinte West. This means that, except for a small bit at the west end that is still in Brighton Township, the Murray Canal is now in Quinte West. Historical references should say Murray Township, but anything after 1998 should say Quinte West.

Here is a more complete record of the scope of the amalgamation:

> Quinte West was formed on January 1, 1998, through the amalgamation of the city of Trenton, the village of Frankford and the townships of Murray and Sidney. Trenton is the largest community and serves as the administrative and commercial centre. In addition to Trenton and Frankford, the district of Quinte West also includes the communities of Barcovan Beach, Batawa, Bayside, Carrying Place, Chatterton, German's Landing, Glen Miller, Glen Ross, Halloway, Johnstown, Lovett, Madoc Junction, Maple View, Mount Zion, Oak Lake, River Valley, Roseland Acres, Spencers Landing, Stockdale, Tuftsville, Twelve O'Clock Point, Wallbridge and Wooler.[2]

Most of Murray Canal is now in Quinte West.

Tug contains displays regarding Murray Canal

Today, it is a pleasant drive down Highway 33 to Carrying Place, and on the south side of the canal, there is a road that provides easy access to much better views of the current road bridge, plus what is left of the pier of the original road bridge and, farther west, the railroad bridge and pier.

At the end of the parking lot, there is an interesting display about the Murray Canal housed in the replica of an old boat. The visitor can see displays inside the boat's cabin, which provide history of the canal including the connection with Sir John A. Macdonald and information about the ships that sailed on the canal.

Continue on down the road past the displays and you will see the old railway bridge sitting on its pier. Fishermen are its main company these days, other than the occasional graffiti artist. Makes you wonder how many people who glide by in their boats have any idea of the history behind this relic.

Turn around at the railway bridge and you will see that a smaller road goes south. This road is what remains of the track bed of the Central Ontario Railway, which stopped running in 1972.[3] The land that held the tracks through Prince Edward County was sold to the county in 1997.[4]

A walking and bicycle trail comes from Trenton to the spot on the north side of the railway bridge, but you have to go to the Highway 33 bridge to cross the canal. Then you can go down this trail all the way to County Road 64, just north of the intersection with Portage Road.

At the western end of the Portage Road, in amongst the new houses, the Millennium Trail begins. Now you are entering Prince Edward County. Check out the website for more information about the route and attractions along the way to Picton.[5]

My Favorite Bike Ride

I recall, as a kid, long Sunday drives into "the County," with the canal being a point of interest. My home in the 1950s and 1960s was a mixed dairy farm at Codrington, but the twenty-year-old farm boy could not get away fast enough when the time came. In the year 2000, genealogy work expanded my knowledge of Prince Edward County, with some hot Sundays spent finding cemeteries and taking pictures of memorials. The Murray Canal was back in my thoughts as part of the path to the County.

More recently, I have had some very pleasant contact with the Murray Canal. In 2010, I moved back to Brighton from downtown Toronto in order to distance myself from big city life and begin to engage in local history in the Brighton area.

Several decades of intense technical support work in south-western Ontario, Calgary, and Toronto brought me to the conviction that I had to get back home to do local history as well as to ensure my own health and well-being. My first home in Brighton was a new town-house on Chapel Street, across from Trinity St. Andrew's United Church. This spacious home served as both comfortable living quarters and a very convenient home office. The basement supported my computer consulting work as well as my growing activities in local history. In this context, "The History Guy of Brighton" was born!

In the spring of 2011, I purchased a bicycle with the idea that I needed to spend more time outside and engage in physical activity. Very quickly, my favorite bike ride became the route down Prince Edward Street, out County Road 64, all the way to Murray Canal and back. This was a perfect route for me since it was mostly flat, so no hills to deal with. Also, it was

The author's bike at Murray Canal

safer than any other route because County Road 64 had been enhanced with bike lanes on both sides.

The swing bridge at the canal became my turn-around point. My habit was to pass over the bridge, if it was not swung for boats, and have a drink of water on the grassy area down the dirt road on the south side of the canal. The picture of glorious late-day sun on the bridge was taken in 2015.

Sometimes, when I arrived at the canal later in the afternoon, there would be boats tied up along the south side of the canal just west of the bridge. If boaters did not arrive at the bridge before five pm, when the bridgetender left for the day, they had to wait until the next morning to go through. Being naturally curious, I would often engage the people on the boats in conversation, usually ending up sharing my knowledge of canal history.

Old bridge at County Road 64 in 2015

Many of the boaters were American, from any different place one might care to mention. One fellow from Florida pointed up the canal and exclaimed that if this kind of facility was in the States, there would be tourist shops and restaurants all along the bank and they would have some entertainment as they waited. For the speaker, it seemed like a legitimate complaint about a wonderful resource gone to waste.

However, as a local historian, I was able to explain that the land was a park and there was always limited development on park land, just to preserve the environment. I'm not sure my American friend was convinced.

Murray Canal Celebrates Its 125th Birthday

The Murray Canal was 125 years old in 2014, and the community held a birthday party on October 18, in the canal park area east of the Highway 33 bridge. There were about one hundred people in attendance, including The History Guy of Brighton. Many folks were dressed in period garb to reflect the historical aspect of the celebration. The crowd was charmed by a couple dressed very authentically as Sir John A. Macdonald and his wife.[6]

The event was organized by the Murray Canal Organization in partnership with History Lives Here Inc., which creates historical videos, led by Peter Lockyer, a well-known historian in Picton. This event was linked to the Macdonald Heritage Trail and the Macdonald Project of Prince Edward County. The idea was to make this an annual event in order to raise awareness of local history in the community.

New Bridge

Even as the 125th anniversary celebration occurred, a lively public debate was ramping up related to plans to replace the swing bridge at County Road 64, which is the modern name for the old Brighton Road. Word circulated that the government was planning to replace the old 1947 bridge with a modern structure. The plans were to include a pedestrian walkway, but the community was shocked to learn that it would only provide one lane of traffic, requiring lights at each end to control traffic, like at construction sites.[7]

People were outraged! How dare they reduce bridge traffic to only one lane? There were weak excuses about cost, but nobody was buying that. Technically speaking, the rational was that a full-sized bridge with two lanes of traffic and a safe pedestrian walkway would require much more substantial replacement of the piers and substructure of the bridge. The main abutments would have to be longer and heavier, and therefore dug much further into the bank than the current structures. The cost of this would bust the budget that had been allocated.

There was a period of very contentious debate on this issue. The Murray Canal District Organization pulled out all the stops, gathering support from local residents as well as businesses and politicians. Of course, any time we are talking about major public infrastructure projects, it becomes political. Politicians have to vote to spend big chunks of public money, and they see their political careers wrapped up in projects like this.

Poster to fight for a full two-lane bridge at County Road 64

Finally, in August of 2016, all the lobbying appeared to have worked. An article on the *InQuinte News* web site stated:

> A Mississauga company has been awarded the contract to design the new two-lane swing bridge across the Murray Canal in Brighton. Trent-Severn Waterway spokesperson Natalie Austin says MMM Group Limited will be paid two-million dollars to do the design work for the structure at County Road 64. It is part of an $8.3-million investment by the federal government in the new bridge, which will have full highway load rating and no restrictions for emergency and service vehicles.[8]

This seemed too good to be true, but on March 22, 2017, a public meeting was held by Parks Canada to discuss the plans. It was attended by residents, business owners, and elected officials for Quinte West, Brighton, and Prince Edward County. "Attendees learned about how construction might impact them, saw preliminary design sketches, and were able to speak to the engineers managing the project. County Road 64, and the Murray Canal and adjacent trails [were] closed at this location to facilitate the construction during the non-navigation season. Residents and visitors [were] asked to follow detour signage."[9]

The prospects for a new bridge looked even brighter when, on one of my bike rides to Murray Canal and back in April, I found a sign that Parks Canada had installed along the road near the bridge. It said:

> Parks Canada Infrastructure Project; Brighton Road Swing Bridge Replacement; The Brighton Road Swing Bridge is to be replaced with a new two lane swing bridge that will have a pedestrian pathway, a full highway load rating eliminating the current restrictions for emergency and service vehicles, and upgraded mechanical and electrical features. The bridge will be fabricated off-site and transported to the site in sections for re-assembly, decreasing the length of closure. The concrete road abutment and centre pier will be rehabilitated to accommodate the new bridge. Construction will take place from mid-September 2017 to mid-May 2018. The replacement of this bridge is part of a $3 billion Parks Canada initiative to support infrastructure work to heritage, visitor, waterway and highway assets located

within national historic sites, national parks, and national marine conservation areas across Canada.[10]

High water problems in 2017

It was looking like this was actually going to happen.

Of course, that summer of 2017 will be etched into the memories of many people who live on the waterways of the province. The water level in Lake Ontario was extremely high, and that was reflected in all associated waterways and shorelines. Here is a picture I took in June of 2017, showing the old bridge at County Road 64 and the unusually high water level in the canal. The old bridge was certainly looking its age, and this high water did not help matters at all. A new bridge was looking like a very good idea.

Right on schedule, the bridge was closed in mid-September 2017 and fencing installed. Sure enough, my normal bike route was interrupted. I could no longer go across the bridge for my drink of water. For a while, I had to be satisfied with stopping near the work site for my refreshment.

Old bridge removed October 2017; preparing for new bridge

On October 20, I took pictures of the work. The bridge had already been removed, and a lot of digging was underway. All around the fenced-in area, there were piles of girders and stone and dirt. I have seldom seen so many machines with long arms on the front with different shaped buckets and digging tools. For someone not accustomed to construction work, this kind of site holds some fascination. Of course, the workers were not much inclined to talk; they were just too busy.

The work continued through the spring of 2018. An email from Parks Canada on May 4, 2018, provided a description of the work being done and the expectations we might have for the function of the bridge:

> The bridge is being replaced with a new two lane swing bridge that will also have a pedestrian sidewalk, a full highway load rating eliminating the current restrictions for emergency and service vehicles, and upgraded mechanical and electrical features. The bridge was fabricated off-site and transported to the site in sections for re-assembly, decreasing the length of closure. The bridge abutments and centre pier are being rehabilitated to accommodate the new bridge.[11]

New bridge in place by May 2018; lots more work to do

Unfortunately, there was no estimated date of completion.

By May of 2018, I could see that the new bridge had been put in place, but it was surrounded by machinery and piles of gravel. There were various delays for whatever reasons, and through the summer of 2018, I wondered if the new bridge was ever going to be ready.

Another email from Parks Canada on July 4, 2018 said:

> The Brighton Road Swing Bridge, located along the Trent-Severn Waterway National Historic Site, will remain closed to vehicle traffic until early September. This closure is to facilitate the replacement of the existing swing bridge with a new, two-lane swing bridge and pedestrian sidewalk. The bridge opening has been delayed due to complications with subcontractors' completion of hydraulics systems and fabrication issues with key mechanical elements. The bridge itself has been fabricated and lifted into place, bridge abutments and centre pier have been completed, the south-east retaining wall constructed, drainage structures have been installed, all electrical duct banks have been placed, and the bridge operator's kiosk has been built. Within the next two weeks, a concrete pad will be poured adjacent to the parking area, the new operator's kiosk will be installed and new concrete stairs leading to the canal will be poured. The span drive cylinder, the main hydraulic component that helps to move the bridge, will be fabricated and ready for installation, in addition to other functional mechanical components for the bridge.[12]

Finally, an email on October 10, 2018, from Parks Canada told us that the bridge would be open on the 15th, although traffic would be limited for a time. Occasional temporary closures would continue as they worked to finished the last jobs to make the bridge functional and safe for the public.[13] Not soon enough, we could enjoy the new bridge.

New bridge open for traffic in October 2018

The bridge was completed around the same time that my recreational habits evolved from bike riding to walking, which kept me closer to the town of Brighton. In any case, I am always happy to check out the canal as I pass over the new bridge on my way to the County.

Sometimes, I deliberately make a long detour on my occasional trips to Trenton by going down to County Road 64, which takes me over both bridges. The image of the glassy-calm canal on a beautiful sunny day that appears on the front cover of this book is the result of one of those drives across the new bridge.

When I am there, looking out over the waters of the canal, I am bemused by the knowledge that, in the 1800s, lobbyists and advocates were so instrumental in motivating the politicians to finally agree to fund the building of the Murray Canal.

Then, more than a century later, the same type of lobbyists and advocates fought to make sure the next generation of bridge would be safe and convenient.

Some things don't change as much as they remain the same.

New bridge is two full lanes and a pedestrian walkway.

NOTES

Chapter 1 Simcoe's Canal

1. S. R. Mealing, "Simcoe, John Graves," *Dictionary of Canadian Biography*, vol. 5, University of Toronto/Université Laval, 2003–, accessed August 12, 2023, http://www.biographi.ca/en/bio/simcoe_john_graves_5E.html.

2. Ibid.

3. Elizabeth Simcoe and J. Ross Robertson, *The Diary of Mrs. John Graves Simcoe*, (Toronto: William Briggs, 1911), 115.

4. "The First Canal Age," Canal and River Trust, last modified January 14, 2022, https://canalrivertrust.org.uk/enjoy-the-waterways/canal-history/the-first-canal-age-canal-history.

5. *Order in Council, York, July 18th, 1796, Volume V. 1792–1796* (Supplementary), (Toronto: Toronto Historical Society, 1931), 197.

6. S. R. Mealing, "Simcoe, John Graves," *Dictionary of Canadian Biography*, vol. 5, University of Toronto/Université Laval, 2003–, accessed August 12, 2023, http://www.biographi.ca/en/bio/simcoe_john_graves_5E.html.

7. "The Carrying Place — Governor Simcoe's Canal," *Kingston Chronicle & Gazette*, Vol. XVI, No. 18, November 1, 1834, article downloaded from Digital Kingston.

8. David W. Smyth, "A Map of the Province of Upper Canada," (London: William Faden, April 12, 1800), https://www.davidrumsey.com/luna/servlet/view/all?res=1&sort=Pub_List_No_InitialSort.

9. "Murray Township" (1878), The Canadian County Atlas Digital Project, accessed August 12, 2023, https://digital.library.mcgill.ca/countyatlas/searchmapframes.php.

10. Colin Powles, "A Construction, Operations and Maintenance History of the Murray Canal" (Unpublished manuscript, Summer 1991), Canadian Parks Service, Ontario Regional Office, typescript, 1, http://danbuchananhistoryguy.com/uploads/1/1/5/0/115043459/murray_canal_colin_powles_report_1991_searchable.pdf.

11. Marion Mikel Calnan, Peggy Dymond Leavey, and Julia Rowe Sager, *Gunshot and Gleanings of the Historic Carrying Place, Bay of Quinte*, (Bloomfield: 7th Town/Ameliasburgh Historical Society, 1987), 14.

12. Asa Weller, at Murray, Sept. 10, 1815, *Department of Finance, War of 1812: Board of Claims for Losses, 1813–1848*, RG 19 E 5 (a), Volume 3757, File 2, Petition 1782, Library and Archives Canada, reel t-1138.

13. Powles Report, 89.

14. Ontario Land Registry Records, "Northumberland County, Murray Township, Concession B, Lot 15," Patent, Book 003, p. 118, digital copy from OnLand.ca.

15. W. R. Topham, *United Lodge No. 29 History 1818–1996*, (Brighton, 1996), 2.

16. Powles Report, 2.

17. "The Carrying Place — Governor Simcoe's Canal," *Kingston Chronicle & Gazette*, Vol. XVI, No. 18, November 1, 1834, article downloaded from Digital Kingston.

18. Ibid.

Chapter 2 Early Surveys and Advocates

1. "The Course of History: Lachine Canal National Historic Site," Government of Canada, last modified November 19, 2022, https://parks.canada.ca/lhn-nhs/qc/canallachine/culture/histoire-history/histoire-history.

2. Jennifer McKendry, "Chronology of the Kingston Architecture," accessed August 12, 2023, http://www.mckendry.net/CHRONOLOGY/chronology.htm.

3. Laura Neilson Bonikowsky, "The Evolution of the Welland Canal," *The Canadian Encyclopedia*, lasted modified March 4, 2015, https://www.thecanadianencyclopedia.ca/en/article/welland-canal-feature.

4. Alan Wilson, "Colborne, John, Baron Seaton," *Dictionary of Canadian Biography*, vol. 9, University of Toronto/Université Laval, 2003–, accessed August 12, 2023, http://www.biographi.ca/en/bio/colborne_john_9E.html.

5. Ibid.

6. "The Carrying Place — Governor Simcoe's Canal," *Kingston Chronicle & Gazette*, Vol. XVI, No. 18, November 1, 1834, article downloaded from Digital Kingston.

7. John Witham, "Baird, Nicol Hugh," *Dictionary of Canadian Biography*, vol. 7, University of Toronto/Université Laval, 2003–, accessed August 12, 2023, http://www.biographi.ca/en/bio/baird_nicol_hugh_7E.html.

8. Colin Powles, "A Construction, Operations and Maintenance History of the Murray Canal" (Unpublished manuscript, Summer 1991), Canadian Parks Service, Ontario Regional Office, typescript, 3, http://danbuchananhistoryguy.com/uploads/1/1/5/0/115043459/murray_canal_colin_powles_report_1991_searchable.pdf.

9. Ibid., 4.

10. "Cornwall Canal," St. Lawrencepiks — Seaway History, accessed August 12, 2023, http://stlawrencepiks.com/seawayhistory/beforeseaway/cornwall/.

11. Fernand Ouellet, "Lambton, John George, 1st Earl of Durham," *Dictionary of Canadian Biography*, vol. 7, University of Toronto/Université Laval, 2003–, accessed August 12, 2023, http://www.biographi.ca/en/bio/lambton_john_george_7E.html.

12. Phillip Buckner, "Thomson, Charles Edward Poulett, 1st Baron Sydenham," *Dictionary of Canadian Biography*, vol. 7, University of Toronto/Université Laval, 2003–, accessed August 12, 2023, http://www.biographi.ca/en/bio/thomson_charles_edward_poulett_7E.html.

13. Ibid.

14. "An Act to Repeal Certain Ordinances Therein Mentioned and to Establish a Board of Works in this Province," *Canada Gazette*, Government of Canada (Province of Canada), No. 8, Regular Issue, November 20, 1841, http://central.bac-lac.gc.ca/.redirect?app=cangaz&id=78&lang=eng.

15. Powles Report, 4.

16. Ibid.

17. Ibid., 5.

18. Ibid.

Chapter 3 Optimism and Lobbying

1. Colin Powles, "A Construction, Operations and Maintenance History of the Murray Canal" (Unpublished manuscript, Summer 1991), Canadian Parks Service, Ontario Regional Office, typescript, 6, http://danbuchananhistoryguy.com/uploads/1/1/5/0/115043459/murray_canal_colin_powles_report_1991_searchable.pdf.

2. "Statutes of Ontario. 1851. Vol. III, 1792–1851, Schedule D. New Townships," Item 11, p. 1801, PDF copy provided by HeinOnline.

3. Dan Buchanan, *The Birth of Brighton Township*, (Brighton: Self-published, 2018), 18–19, http://danbuchananhistoryguy.com/uploads/1/1/5/0/115043459/the_birth_of_brighton_township.pdf.

4. Alex Begg, "Murray Canal," *Brighton Sentinel*, 1853 quoted in Wilmot M. Tobey, *The Tobey Book*, ed. Barbara Nyland, (Unpublished manuscript, July 1975), typescript, 289, http://danbuchananhistoryguy.com/uploads/1/1/5/0/115043459/the_tobey_book.pdf.

5. "Gilmour Lumber Company," Wikipedia, last modified November 1, 2022, https://en.wikipedia.org/wiki/Gilmour_Lumber_Company.

6. Powles Report, 6.

7. Alex Begg, *Brighton Sentinel*, April 8, 1853, quoted in Wilmot M. Tobey, *The Tobey Book*, Barbara Nyland, ed., (Unpublished manuscript, July 1975), typescript, 493, http://danbuchananhistoryguy.com/uploads/1/1/5/0/115043459/the_tobey_book.pdf.

8. Monroe Lawson, *Brighton Ensign*, July 21, 1931, quoted in Wilmot M. Tobey, *The Tobey Book*, Barbara Nyland, ed., (Unpublished manuscript, July 1975), typescript, 265–266, http://danbuchananhistoryguy.com/uploads/1/1/5/0/115043459/the_tobey_book.pdf.

9. Samuel Keefer, *Brighton Sentinel*, August 12, 1853, quoted in Wilmot M. Tobey, *The Tobey Book*, Barbara Nyland, ed., (Unpublished manuscript, July 1975), typescript, 492–493, http://danbuchananhistoryguy.com/uploads/1/1/5/0/115043459/the_tobey_book.pdf.

10. Powles Report, 7.

11. Capt. William Quick, *Brighton Sentinel*, 1853, quoted in Wilmot M. Tobey, *The Tobey Book*, Barbara Nyland, ed., (Unpublished manuscript, July 1975), typescript, 259, http://danbuchananhistoryguy.com/uploads/1/1/5/0/115043459/the_tobey_book.pdf.

12. Colonel McDougall, Adjutant General of Militia, House of Commons Committee, 1865, quoted in Wilmot M. Tobey, *The Tobey Book*, Barbara Nyland, ed., (Unpublished manuscript, July 1975), typescript, 399, http://danbuchananhistoryguy.com/uploads/1/1/5/0/115043459/the_tobey_book.pdf.

13. Montreal Board of trade, 1865, quoted in in Wilmot M. Tobey, *The Tobey Book*, Barbara Nyland, ed., (Unpublished manuscript, July 1975), typescript, 494, http://danbuchananhistoryguy.com/uploads/1/1/5/0/115043459/the_tobey_book.pdf.

Chapter 4 In the Halls of Parliament

1. Marion Mikel Calnan, Peggy Dymond Leavey, and Julia Rowe Sager, *Gunshot and Gleanings of the Historic Carrying Place, Bay of Quinte*, (Bloomfield: 7th Town/Ameliasburgh Historical Society, 1987), 199.

2. Ibid.

3. Report from Committee, 1866, quoted in Wilmot M. Tobey, *The Tobey Book*, Barbara Nyland, ed., (Unpublished manuscript, July 1975), typescript, 495, http://danbuchananhistoryguy.com/uploads/1/1/5/0/115043459/the_tobey_book.pdf.

4. Colin Powles, "A Construction, Operations and Maintenance History of the Murray Canal" (Unpublished manuscript, Summer 1991), Canadian Parks Service, Ontario Regional Office, typescript, 8, http://danbuchananhistoryguy.com/uploads/1/1/5/0/115043459/murray_canal_colin_powles_report_1991_searchable.pdf.

5. Ibid.

6. Ibid., 9.

7. Ibid., 10–11.

8. Eileen Argyris, *How Firm a Foundation:* A History of the Township of Cramahe and the Village of Colborne, (Canada: Millennium Committee of the Village of Colborne and the Township of Cramahe, 2000), https://heritagecramahe.ca/keeler-family/.

9. Ibid.

10. Ibid.

11. "Has Shovel Used To Open Work On Murray Canal," from a Colborne newspaper, date unknown, digital copy provided by Lenna Broatch, March 23, 2023. See Appendix C.

12. Powles Report, 11.

13. Ibid.

14. Ibid., 12.

15. J. K. Johnson and P. B. Waite, "Macdonald, Sir John Alexander," *Dictionary of Canadian Biography*, vol. 12, University of Toronto/Université Laval, 2003–, accessed August 12, 2023, http://www.biographi.ca/en/bio/macdonald_john_alexander_12E.html.

16. Powles Report, 12.

17. Ibid.

18. Ibid.

19. Calnan, Leavey, and Sager, *Gunshot and Gleanings*, 199.

20. John P. Heisler, "The Canals of Canada," Department of Indian Affairs and Northern Development, Ottawa, *Canada Historic Sites: Occasional Papers in Archaeology and History*, no. 8 (1973): 135, http://parkscanadahistory.com/series/chs/8/chs8-1a.htm.

21. Ibid., 129.

22. Ibid., 148.

Chapter 5 The Final Push

1. Colin Powles, "A Construction, Operations and Maintenance History of the Murray Canal" (Unpublished manuscript, Summer 1991), Canadian Parks Service, Ontario Regional Office, typescript, 13, http://danbuchananhistoryguy.com/uploads/1/1/5/0/115043459/murray_canal_colin_powles_report_1991_searchable.pdf.

2. Ibid.

3. Marc Seguin, *For Want of a Lighthouse: Building the Lighthouses of Eastern Lake Ontario 1828–1914*, (Bloomington: Trafford Publishing, 2015), 272.

4. Powles Report, 13–14.

5. Ibid., 14.

6. Ibid.

7. Ibid., 14–15.

8. Ibid., 15.

9. Ibid.

10. Seguin, *For Want of a Lighthouse*, 278.

11. Powles Report, 15–16.

12. Ibid., 16.

13. Ibid.

14. Eileen Argyris, *How Firm a Foundation:* A History of the Township of Cramahe and the Village of Colborne, (Canada: Millennium Committee of the Village of Colborne and the Township of Cramahe, 2000), https://heritagecramahe.ca/keeler-family/.

15. Powles Report, 16.

16. Ibid.

17. Ibid.

18. "Thomas David 'Tom' Rubidge," www.treesbydan.com.

19. Powles Report, 16–17.

20. Order in Council, Privy Council Office, Series A-1-a, Volume 415, Access Code 90, Library and Archives Canada, reel C-3338.

21. Seguin, *For Want of a Lighthouse*, 350–351.

22. Powles Report, 17.

23. Order in Council, Privy Council Office, Series A-1-a, Volume 415, Access Code 90, Library and Archives Canada, reel C-3338.

24. Powles Report, 18.

25. Ibid., 19.

Chapter 6 Contracts and Opening the Works

1. Colin Powles, "A Construction, Operations and Maintenance History of the Murray Canal" (Unpublished manuscript, Summer 1991), Canadian Parks Service, Ontario Regional Office, typescript, 20, http://danbuchananhistoryguy.com/uploads/1/1/5/0/115043459/murray_canal_colin_powles_report_1991_searchable.pdf.

2. Ibid.

3. Ibid.

4. Ibid., 21.

5. Ibid.

6. "Approval of Two Land Valuators," Orders in Council, Privy Council Office, Series A-1-a, 1882–1734, Volume 419, Library and Archives Canada, reel C-3340.

7. "Surveyed Murray Canal Re-visits Brighton," *Colborne Express*, July 14, 1938, http://images.ourontario.ca/Partners/CTPL/CTPL002887662pf_0001.pdf.

8. Powles Report, 21–22.

9. Ontario Land Registry Records, "Northumberland County, Murray Township, Concession C, Lot 21," #3039, Book 002, p. 241, digital copy obtained Nov. 21, 2020, from OnLand.ca.

10. Ontario Land Registry Records, "Northumberland County, Murray Township, Concession C, Lot 18," #3040, Book 002, p. 178, digital copy obtained Nov. 21, 2020, from OnLand.ca.

11. Powles Report, 22.

12. Ibid.

13. "Has Shovel Used To Open Work On Murray Canal," from a Colborne newspaper, date unknown, digital copy provided by Lenna Broatch, March 23, 2023. See Appendix C.

14. Powles Report, 22.

15. Ibid., 22–23.

16. Ibid., 23.

17. Ibid.

18. Ibid.

19. Ibid.

20. Ibid.

21. Ibid., 23–24.

Chapter 7 Building the Canal

1. Colin Powles, "A Construction, Operations and Maintenance History of the Murray Canal" (Unpublished manuscript, Summer 1991), Canadian Parks Service, Ontario Regional Office, typescript, 24, http://danbuchananhistoryguy.com/uploads/1/1/5/0/115043459/murray_canal_colin_powles_report_1991_searchable.pdf.

2. James Plomer and Alan R. Capon, *Desperate Venture: The Central Ontario Railway*, (Belleville: Mika Publishing Company, 1979).

3. Powles Report, 24.

4. Ibid.

5. Ibid.

6. Ibid.

7. Madelein "Peggy" Muntz, *John Laing Weller C.E., M.E.I.C.: "The Man Who Gets Things Done,"* (St. Catharines: Vanwell Publishing Limited, 2007), 11.

8. Powles Report, 25.

9. Ibid.

10. Ibid.

11. Ibid.

12. Ontario Land Registry Records, "Northumberland County, Murray Township, Concession C, Lot 17," Patent, Book 002, p. 160, digital copy obtained Nov. 21, 2020, from OnLand.ca.

13. Powles Report, 21.

14. "Dredging," *Weekly Intelligencer*, Thursday, September 11, 1890, https://archive.org/details/intelligencer-weekly-september-1890/page/n15/mode/2up?q=weddell.

15. "Release ½ Security Deposit," Orders in Council, Privy Council Office, Series A-1-a, Vol. 464, Item #632, Library and Archives Canada, reel C-3372.

16. Powles Report, 25.

17. Ibid., 25–26.

18. Ibid., 8.

19. Ibid., 20.

20. Ibid., 26.

21. Ibid.

22. Ibid.

23. Ibid., 26–27.

24. Ibid., 27.

25. Ibid.

26. Ibid., 28.

27. Ibid.

28. Peter E. Paul Dembski, "Cochrane, Edward," *Dictionary of Canadian Biography,* vol. 13, University of Toronto/Université Laval, 2003–, accessed August 12, 2023, http://www.biographi.ca/en/bio/cochrane_edward_13E.html.

29. Powles Report, 28–29.

30. "Trenton Bridge and Engine Works," *Trenton Directory* 1888, 283–284.

31. "Robert Weddell Sr. (1821–1898)," www.treesbydan.com.

32. Plomer and Capon, *Desperate Venture,* 125.

33. "Thriving Industries," *Belleville Daily Intelligencer,* May 19, 1888, https://archive.org/details/intelligencer-may-1888/page/n63/mode/2up?q=%22robert+weddell%22.

34. Marion Mikel Calnan, Peggy Dymond Leavey, and Julia Rowe Sager, *Gunshot and Gleanings of the Historic Carrying Place, Bay of Quinte,* (Bloomfield: 7th Town/Ameliasburgh Historical Society, 1987), 91.

35. "Dredging," *Weekly Intelligencer,* Thursday, September 11, 1890, https://archive.org/details/intelligencer-august-1888/page/n65/mode/2up?q=robert+weddell.

Chapter 8 Demonstration and More Work

1. "Murray Canal," Hastings County Historical Society, Community Archives of Belleville and Hastings County, 1886, donated by Roy Bruce, https://discover.cabhc.ca/murray-canal.

2. Colin Powles, "A Construction, Operations and Maintenance History of the Murray Canal" (Unpublished manuscript, Summer 1991), Canadian Parks Service, Ontario Regional Office, typescript, 27 http://danbuchananhistoryguy.com/uploads/1/1/5/0/115043459/murray_canal_colin_powles_report_1991_searchable.pdf.

3. Ibid., 27–28.

4. "Release Security Deposit," Orders in Council, Privy Council Office, Series A-1-a, Vol. 464, Order in Council No. 1885-0632, Library and Archives Canada, reel C-3372.

5. "Release Balance of Security Deposit," Orders in Council, Privy Council Office, Series A-1-a, Vol. 496, Order in Council No. 1887-0282, Library and Archives Canada, reel C-3384.

6. Powles Report, 29.

7. Ibid.

8. Ibid.

9. Ibid.

10. Ibid.

11. Ibid.

12. Ibid.

13. Donna Fano, "Bridging Belleville and Prince Edward County," *Outlook,* Volume 22 Number 1, Issue 312 Jan 2017, Hastings County Historical Society, 4, https://web.archive.org/web/20210626222216/https://hastingshistory.ca/News/news.inc.php?ID=56&command=miniViewArticle&lang=EN&s=0 .

Chapter 9 "Canalis" Tours the Works

1. "The Murray Canal," *Belleville Daily Intelligencer*, Thursday, August 23, 1888, https://archive.org/details/intelligencer-august-1888/page/n65/mode/2up?q=%22robert+weddell%22.

2. Ibid.

3. Ibid.

4. Ibid.

5. Ibid.

6. Ibid.

7. Ibid.

8. Ibid.

9. Ibid.

10. Ibid.

11. Ibid.

Chapter 10 Final Work and Opening

1. Colin Powles, "A Construction, Operations and Maintenance History of the Murray Canal" (Unpublished manuscript, Summer 1991), Canadian Parks Service, Ontario Regional Office, typescript, 31, http://danbuchananhistoryguy.com/uploads/1/1/5/0/115043459/murray_canal_colin_powles_report_1991_searchable.pdf.

2. "Special Warrant," Orders in Council, Privy Council Office, Series A-1-a, Vol. 528, Order in Council No. 1888-2086, Library and Archives Canada, reel C-3395.

3. "Pay for House," Orders in Council, RG2, Privy Council Office, Series A-1-a, Vol. 529, Order in Council No. 1888-2241, Library and Archives Canada, reel C-3396.

4. Powles Report, 31.

5. Ibid.

6. Ibid.

7. "Thomas Phillips Keeler (1852–c1901)," #39928, www.treesbydan.com.

8. George Taylor, letter to Prime Minister Sir John A. Macdonald, June 17, 1889, *Sir John A. Macdonald Papers*, Volume 135, Item No. 543: 55871, Library and Archives Canada.

9. "T.P. Keeler Appointment," Order in Council, Item #1889-1530, Library and Archives Canada.

10. Powles Report, 32.

11. "Looking Backward, 50 Years Ago, August 16, 1889," *Ontario Intelligencer*, August 16, 1939, Internet Archive, https://archive.org/details/intelligencer-august-1939/page/n143/mode/2up. ,

12. "Alexander Forbes (1842–1916)," #115552, www.treesbydan.com.

13. "Mowry Recognized," Orders in Council, RG2, Privy Council Office, Series A-1-a, Vol. 558, Order in Council No. 1890-0996, Library and Archives Canada, reel C-3407.

14. "John David Silcox (1847–1919)," #64871, www.treesbydan.com.

15. Powles Report, 20.

16. "Collingwood Shipbuilding," Wikimili, last modified October 15, 2022, https://wikimili.com/en/Collingwood_Shipbuilding.

17. "Willett McConnell Platt (1820–1902)," #64857, www.treesbydan.com.

18. John David Silcox (1847-1919), #64871, www.treesbydan.com.

19. Ontario Land Registry Records, "Northumberland County, Brighton Village Lot 1, Main St. North Side," Book 004, p. 271, digital copy from OnLand.Ca.

20. "Mary Elizabeth Platt (1862–1900)," #64863, www.treesbydan.com.

21. "Rules Apply to Murray Canal," Orders in Council, RG2, Privy Council Office, Series A-1-a, Vol. 558, Order in Council No. 1890-1217, Library and Archives Canada, reel C-3407.

22. "Fixing the Rates," Orders in Council, RG2, Privy Council Office, Series A-1-a, Vol. 558, Order in Council No. 1890-1216, Library and Archives Canada, reel C-3407.

23. Marc Seguin, *For Want of a Lighthouse: Building the Lighthouses of Eastern Lake Ontario 1828–1914*, (Bloomington: Trafford Publishing, 2015), 321–322.

24. Ibid., 322.

Chapter 11 Operations and Maintenance

1. Colin Powles, "A Construction, Operations and Maintenance History of the Murray Canal" (Unpublished manuscript, Summer 1991), Canadian Parks Service, Ontario Regional Office, typescript, 32, http://danbuchananhistoryguy.com/uploads/1/1/5/0/115043459/murray_canal_colin_powles_report_1991_searchable.pdf.

2. Ibid.

3. Ibid., 35.

4. Ibid., 33.

5. Ibid., 37.

6. Ibid., 60–61.

7. Ibid., 62.

8. Ibid., 33.

9. Ibid., 34.

10. Ibid., 33.

11. J. K. Johnson and P. B. Waite, "Macdonald, Sir John Alexander," *Dictionary of Canadian Biography,* vol. 12, University of Toronto/Université Laval, 2003–, accessed August 12, 2023, http://www.biographi.ca/en/bio/macdonald_john_alexander_12E.html.

12. "The Murray Canal and Its Benefits to Trade and Travel," *Daily Intelligencer,* Belleville, July 2, 1891, Community Archives Belleville and Hastings County, https://discover.cabhc.ca/uploads/r/community-archives-of-belleville-and-hastings-county/4/2/e/42e8c015f542a58d0e27570108697459cec1f8527463a98d8da6c05e0a3bb4d3/HCHS-03-02-10.pdf.

13. "Murray Canal and Approaches, including Aids to Navigation," *Canada Gazette*, 1867–1946 (Dominion of Canada), vol. 25, no. 18 (October 31, 1891): 28–29, Library and Archives Canada, https://recherche-collection-search.bac-lac.gc.ca/eng/home/record?app=cangaz&IdNumber=3223.

14. Powles Report, 48.

15. Ibid., 43.

16. "Payment in Full," Orders in Council, RG2, Privy Council Office, Series A-1-a, Vol. 629, Order in Council No. 1893-2180, Library and Archives Canada, reel C-3613.

17. Marc Seguin, *For Want of a Lighthouse: Building the Lighthouses of Eastern Lake Ontario 1828–1914*, (Bloomington: Trafford Publishing, 2015), 326.

18. Powles Report, 62.

19. Ibid.

20. Ibid., 63.

Chapter 12 Fixing and Changing

1. Marc Seguin, *For Want of a Lighthouse: Building the Lighthouses of Eastern Lake Ontario 1828–1914*, (Bloomington: Trafford Publishing, 2015), 323.

2. Ibid.

3. "COR Tracks North of Canal," Orders in Council, RG2, Privy Council Office, Series A-1-a, Vol. 777, Order in Council No. 1899-0535, Library and Archives Canada, reel C-3770.

4. "T.P. Keeler Appointed Superintendent," Orders in Council, RG2, Privy Council Office, Series A-1-a, Vol. 789, Order in Council No. 1899-244, Library and Archives Canada, reel C-3775.

5. "Investigation of Charges," Orders in Council, RG2, Privy Council Office, Series A-1-a, Vol. 792, Order in Council No. 1900-0125, Library and Archives Canada, reel C-3776.

6. "Dismissal of Superintendent," Orders in Council, RG2, Privy Council Office, Series A-1-a, Vol. 797, Order in Council No. 1900-0757, Library and Archives Canada, reel C-3779.

7. Ibid.

8. Colin Powles, "A Construction, Operations and Maintenance History of the Murray Canal" (Unpublished manuscript, Summer 1991), Canadian Parks Service, Ontario Regional Office, typescript, 32, http://danbuchananhistoryguy.com/uploads/1/1/5/0/115043459/murray_canal_colin_powles_report_1991_searchable.pdf.

9. Ibid., 63–64.

10. Ibid., 64.

11. Ibid., 61.

12. Ibid.

13. Ibid., 61–62.

14. "Accept Weddell Tender for Bridges," Orders in Council, RG2, Privy Council Office, Series A-1-a, Vol. 916, Order in Council No. 1906-1903, Library and Archives Canada.

15. Ibid.

16. "Extend Contract for Piers," Orders in Council, RG2, Privy Council Office, Series A-1-a, Vol. 942, Order in Council No. 1907-2872, Library and Archives Canada.

17. Powles Report, 41.

18. "COR Lease," Orders in Council, RG2, Privy Council Office, Series A-1-a, Vol. 960, Order in Council No. 1908-2102, Library and Archives Canada.

19. Powles Report, 64.

20. Ibid.

21. Ibid.

22. Ibid., 65.

23. Ibid., 64–65.

24. Ibid., 65.

25. Ibid.

26. Ibid., 66.

27. Ibid.

28. Ibid.

29. Ibid.

30. Ibid., 33.

31. Ibid., 34.

32. Seguin, *For Want of a Lighthouse*, 351.

33. Powles Report, 48.

34. Ibid., 34.

35. Ibid., 66–67.

36. Ibid., 34.

37. Ibid., 48–49.

38. Ibid., 43.

39. Ibid., 36–37.

40. Ibid., 43.

Chapter 13 Problems with Bridges

1. Colin Powles, "A Construction, Operations and Maintenance History of the Murray Canal" (Unpublished manuscript, Summer 1991), Canadian Parks Service, Ontario Regional Office, typescript, 44, http://danbuchananhistoryguy.com/uploads/1/1/5/0/115043459/murray_canal_colin_powles_report_1991_searchable.pdf.

2. Ibid.

3. Ibid., 49.

4. Ibid., 34–35.

5. Ibid., 35.

6. Ibid., 44.

7. Ibid., 44–45.

8. Ibid., 45.

9. Ibid.

10. Ibid.

11. Ibid.

12. Seventh Town Historical Society, "Chapter 6: The Carrying Place," in *7th Town/Ameliasburgh Township, Past & Present*, (Milton: Global Heritage Press, 1999), 71 & 73.

13. Powles Report, 45.

14. Ibid., 46.

15. Ibid.

16. Ibid.

17. Ibid., 46–47.

18. Ibid., 47.

19. Ibid.

20. Ibid.

21. Ibid.

22. Ibid.

23. Ibid.

24. Ibid., 49.

25. Ibid.

26. Ibid.

27. Ibid.

28. Ibid., 49–50.

29. Ibid., 50.

30. Ibid.

Chapter 14 People of the Canal

1. Ontario Land Registry Records, "Northumberland County, Murray Township, Carrying Place Lot 13," #3167, Book 01, p. 300, digital copy from OnLand.ca.

2. "Louis Latour (1839–1895)," #112070, www.treesbydan.com.

3. Ontario Land Registry Records, "Carrying Place Lot 13," #3317, Book 001: 300, digital copy from OnLand.ca.

4. Ontario Land Registry Records, "Carrying Place Lot 10," #3388, Book 01: 120, digital copy from OnLand.ca.

5. Ontario Land Registry Records, "Carrying Place Lot 12," #D485, Book 001: 235, digital copy from OnLand.ca.

6. "Benjamin Rowe (1812–1897)," #89027, www.treesbydan.com.

7. Ontario Land Registry Records, "Carrying Place Lot 9," #3172, Book 001, p. 086, digital copy from OnLand.ca.

8. Ontario Land Registry Records, "Carrying Place Lot 8," #3178, Book 001, p. 076, digital copy from OnLand.ca.

9. Ontario Land Registry Records, "Carrying Place Lot 7," #3178, Book 001, p. 067, digital copy from OnLand.ca.

10. Ontario Land Registry Records, "Carrying Place Lot 6," #3120, Book 001, p. 056, digital copy from OnLand.ca.

11. Thomas A. Porter (1837-1918), #86458, www.treesbydan.com.

12. Ontario Land Registry Records, "Concession C, Lot 8," #3181, Book 002, p. 012, digital copy from OnLand.ca.

13. Ontario Land Registry Records, "Concession C, Lot 9," #3181, Book 002, p. 023, digital copy from OnLand.ca.

14. Ontario Land Registry Records, "Concession C, Lot 8," Patent, Book 002, p. 012, digital copy from OnLand.ca.

15. Ibid., #3186.

16. Ontario Land Registry Records, "Concession C, Lot 10," #556, Book 002, p. 047, digital copy from OnLand.ca.

17. Ontario Land Registry Records, "Concession C, Lot 11," #540, Book 002, p. 063, digital copy from OnLand.ca.

18. Ibid., Patent.

19. Ibid., #D14.

20. Ibid., #D15.

21. Ontario Land Registry Records, "Concession C, Lot 13," #3085, Book 002, p. 103, digital copy from OnLand.ca.

22. Colin Powles, "A Construction, Operations and Maintenance History of the Murray Canal" (Unpublished manuscript, Summer 1991), Canadian Parks Service, Ontario Regional Office, typescript, 25, http://danbuchananhistoryguy.com/uploads/1/1/5/0/115043459/murray_canal_colin_powles_report_1991_searchable.pdf.

23. "Peter Gould (1824–1891)," #111993, www.treesbydan.com.

24. Ibid.

25. Ontario Land Registry Records, "Concession C, Lot 13," #3089, Book 002, p. 103, digital copy from OnLand.ca.

26. Ontario Land Registry Records, "Concession C, Lot 14," #3110, Book 002, p. 120, digital copy from OnLand.ca.

27. Ontario Land Registry Records, "Concession C, Lot 15," #3082, Book 002, p. 134, digital copy from OnLand.ca.

28. Ontario Land Registry Records, "Concession C, Lot 16," #3052, Book 002, p. 146, digital copy from OnLand.ca.

29. "Abraham Stoneburgh (1796–1873)," #33769, www.treesbydan.com.

30. Ontario Land Registry Records, "Concession C, Lot 17," #3041, Book 002, p. 161, digital copy from OnLand.ca.

31. Ontario Land Registry Records, "Concession C, Lot 17," #2745, Book 002, p. 161, digital copy from OnLand.ca.

32. "Emily Jane Williams (1850–1902)," #11153, www.treesbydan.com.

33. Ontario Land Registry Records, Concession C, Lot 17," #3053, Book 002, p. 161, digital copy from OnLand.ca.

34. "Mary McKenzie (1805–c1895)," #111999, www.treesbydan.com.

35. Ibid.

36. Ontario Land Registry Records, "Concession C, Lot 18," #B161, Book 002, p. 178, digital copy from OnLand.ca.

37. Ibid., #3040.

38. Ontario Land Registry Land Records, "Concession C, Lot 19," #3081, Book 002, p. 201, digital copy from OnLand.ca.

39. Ontario Land Registry Records, "Concession C, Lot 20," #3030, Book 002, p. 215, digital copy from OnLand.ca.

40. Ibid., #3038.

41. Ibid., #3032.

42. "John Wilson (c1790– ?)," #112055, www.treesbydan.com.

43. Ontario Land Registry Records, "Concession C, Lot 21," #3039, Book 002, p. 241, digital copy from OnLand.ca.

44. Ontario Land Registry Records, Concession C, Lot 22, #3031, Book 002, p. 271, digital copy from OnLand.ca.

45. Ibid., #D246.

46. Ontario Land Registry Records, Concession B, Lot 13, #3397, Book 003, p. 087, digital copy from OnLand.ca.

47. Ibid.

48. "Samuel May (1811–1898)," #51897, www.treesbydan.com.

49. "Samuel May (1821–1903)," #86536, www.treesdbydan.com.

50. Ontario Land Registry Records, Concession B, Lot 14, #3146, Book 003, p. 102, digital copy from OnLand.ca.

51. Ontario Land Registry Records, Concession B, Lot 15, #3145, Book 003, p. 119, digital copy from OnLand.ca.

52. Ibid., #530.

53. Ibid., #528.

54. Ontario Land Registry Records, Concession B, Lot 16, #3115, Book 003, p. 148, digital copy from OnLand.ca.

55. Ibid., #2841.

56. Ontario Land Registry Records, Concession B, Lot 17, #3093, Book 003, p. 170, digital copy from OnLand.ca.

57. Ontario Land Registry Records, Concession B, Lot 18, #3088, Book 003, p. 186, digital copy from OnLand.ca.

58. "Land Expropriation Jonathan McMaster," Orders in Council, RG2, Privy Council Office, Series A-1-a, Vol. 421, Order in Council No. 1882-2015, Library and Archives Canada, reel C-3340.

59. "John McMaster (1818-1888)," #74318, www.treesbydan.com.

60. Ontario Land Registry Records, "Concession C, Lot 22," #A65, Book 002, p. 22, digital copy from OnLand.ca.

61. Ontario Land Registry Records, "Brighton Twp., Concession C, Lot 23," #A61, p. 094, digital copy from OnLand.ca.

62. Ontario Land Registry Records, "Brighton Twp., Concession C, Lot 24," #A167, p. 111, digital copy from OnLand.ca.

63. Ontario Land Registry Records, "Brighton Twp., Concession C, Lot 25," #A167, p. 120, digital copy from OnLand.ca.

64. Ontario Land Registry Records, "Brighton Twp., Concession C, Lot 26," #3569, p. 140, digital copy from OnLand.ca.

65. Ontario Land Registry Records, "Brighton Twp., Concession C, Lot 27," #3545, p. 184, digital copy from OnLand.ca.

66. Ontario Land Registry Records, "Brighton Twp., Concession C, Lot 28," #3556, p. 196, digital copy from OnLand.ca.

67. Marion Mikel Calnan, Peggy Dymond Leavey, and Julia Rowe Sager, *Gunshot and Gleanings of the Historic Carrying Place, Bay of Quinte*, (Bloomfield: 7th Town/Ameliasburgh Historical Society, 1987), 88.

68. "John Hanna (1861-1944)," #13145, www.treesbydan.com.

69. Ibid.

70. "Harold Grant Hanna (1898–1965)," #13267, www.treesbydan.com.
    Powles Report, 53.

71. "William Thomas Horsley (1898–1962)," #88212, www.treesbydan.com.

72. "John Robert Harvey (1917–2000)," #39031, www.treesbydan.com.

73. Powles Report, 54.

74. "Allen Frederick Alyea (1914–1999)," #9919, www.treesbydan.com.

75. Ontario Land Registry Land Records, "Murray Twp., Concession C, Lot 7," #L3678, Book 002, p. 007, digital copy from OnLand.ca.

76. Ontario Land Registry Land Records, "Concession C, Lot 8," #L3678, Book 002, p. 012, digital copy from OnLand.ca.

77. "Allen Edwin 'Marcus' Lovett (1852–1932)," #76332, www.treesbydan.com.

78. Ontario Land Registry Land Records, Concession C, Lot 21, #3039, digital copy from OnLand.ca, Book 002, p. 241.

79. Powles Report, 51.

80. "Purchase Goodrich House," Orders in Council, Privy Council Office, Series A-1-a, Vol. 847, Order in Council No. 1902-1786, Library and Archives Canada.

81. Powles Report, 51.

82. "Payment for House," Orders in Council, Privy Council Office, Series A-1-a, Vol. 872, Order in Council No. 1904-0642, Library and Archives Canada.

83. Ibid.

84. Powles Report, 51–52.

85. Ibid., 52.

86. Ibid., 53.

87. Ibid.

88. Ibid., 54–55.

Chapter 15 The Murray Canal Today

1. Orland French, ed., *Heritage Atlas of Hastings County*, (Belleville: Wallbridge House Publishing, 2006), 201.

2. "Quinte West," Wikipedia, last modified June 2023, 21, https://en.wikipedia.org/wiki/Quinte_West.

3. James Plomer and Alan R. Capon, *Desperate Venture: The Central Ontario Railway*, (Belleville: Mika Publishing Company, 1979), 202.

4. "Millennium Trial," Visit the County, accessed August 12, 2023, https://www.visitthecounty.com/business_listings/millennium-trail/.

5. Ibid.

6. "Murray Canal Celebrates 125th Anniversary," History Lives Here, Peter Lockyer, last modified October 18, 2014, https://historyliveshere.ca/murray-canal-celebrates-125th-anniversary/.

7. "Single Lane Bridge," Murray Canal District Organization, https://murraycanaldistrict.ca/opportunities/single-lane-bridge-delayed-need-funds/.

8. County Road 64 Bridge Notice, *InQuinte News*, last modified August 23, 2016, https://inquinte.ca/story/design-contract-awarded-for-murray-canal-swing-bridge1.

9. "Brighton Swing Bridge," Parks Canada (Government of Canada), last modified April 18, 2019, https://www.pc.gc.ca/en/lhn-nhs/on/trentsevern/visit/infrastructure/quinte/pont-tournant-chemin-brighton-road-swing-bridge.

10. *Parks Canada Sign*, photo taken by author, April 27, 2018, digital copy of text provided here.

11. "Community Update: Replacement of Brighton Road Swing Bridge," Parks Canada (Government of Canada), last modified May 04, 2018, https://parks.canada.ca/lhn-nhs/on/trentsevern/visit/infrastructure/quinte/~/link.aspx?_id=7058742E1E8C42BD856553743298A38E&_z=z

12. "Community Update: Work on the Swing Bridge to Continue into September," Parks Canada (Government of Canada), last modified July 4, 2018, https://parks.canada.ca/lhn-nhs/on/trentsevern/visit/infrastructure/quinte/pont-tournant-chemin-brighton-road-swing-bridge/julliet-bulletin-collectivite-july-community.

13. "Info-Work: Construction Update," Parks Canada (Government of Canada), last modified October 5, 2018, https://parks.canada.ca/lhn-nhs/on/trentsevern/visit/infrastructure/quinte/pont-tournant-chemin-brighton-road-swing-bridge/2018-octobre-info-travaux-work-october.

APPENDIX A:
Order in Council Approving the Route

On a Report date 20th May 1882 from the Minister of Railways and Canals, stating that in pursuance of a Vote of Parliament passed in the Session of 1880-81 towards the execution of works involved in the construction of a Canal across the isthmus dividing the waters of the Bay of Quinte from Lake Ontario, instructions were given to the Chief Engineer of Government Canals to the end that a survey and estimate might be made by a competent engineer for the determination of the most feasible line and terminal point on Lake Ontario for such a Canal.

That such survey and comparative estimate have been duly made and submitted by Mr. Thomas Rubidge C.E. in a communication dated the 1st of February last.

That from such report it appears that of the points examined with a view to their adaptability as a port of entrance from Lake Ontario, specifically the points known respectively as Weller's Bay and Presqu'ile, Presqu'ile is by far the most commodious and best harbor on the Coast, having excellent anchorage and enabling a large number of vessels to be land-locked, secure from all winds — further, that the route having this harbor as its western terminus is the one best adapted to the requirement of an extended river navigation.

That the total length of the Canal proper via Weller's Bay is 4 miles, 660 feet, while the length via Presqu'ile is 6 miles, 660 feet, or a difference in favor of Weller's Bay in point of length of 2 miles, this difference is not however held to be of weight inasmuch as the excess by the Presqu'ile route lies through the land-locked harbor of that place.

That the cost of the two several routes estimated as follows, the calculations being based a Scheme for a Canal of a depth of 11 feet at lowest water and without locks, dredging included –

Feet Wide at Bottom	Presqu'ile	Weller's Bay
150	$974,000	$1,472,000
100	792,000	1,302,000
80	721,000	1,229,000

That with reference to the route via Weller's Bay it appears to be a fact from the evidence obtained that the channel across the bar at that place is of a shifting character and that its position or direction is not to be depended upon after a storm — also that while the harbor affords good holding ground and deep water it gives no shelter from the heavy sea rolling in from the Lake before south-westerly or westerly gales.

The Minister therefore recommends that authority be given for the adoption of the route having its western terminal point at Presqu'ile and for the commencement of the works contemplated in the special vote of $200,000 for the Murray Canal — the width to be 80 feet at bottom and the cost of such canal being estimated at $721,000.

The Committee concur in the foregoing report and recommendations of the Minister of Railways and Canals and submit the same for Your Excellency's approval.

Signed: A. W. McLelan
Approved: 23rd May 1882

Source: *Murray Canal - Min R. and C. [Minister of Railways and Canals], 1882/05/20, rec [recommends] adoption of route having its western terminal point at Presqu'ile and for works contemplated in vote of $200,000, width 80 feet at bottom* Order in Council 1882-1081, *RG2, Privy Council Office, Series A-1-a, Volume 415, Access Code 90, Library and Archives Canada, reel C-3338, https://recherche-collection-search.bac-lac.gc.ca/eng/home/record?app=ordincou&IdNumber=20991&q=Order%20in%20 Council%201882-1081.*

APPENDIX B:
"Canalis" Tours the Works in 1888

Article in the *Belleville Daily Intelligencer* on Thursday, August 23 1888.

Daily Intelligencer
Belleville, Thursday, Aug. 23.

The Murray Canal
An interesting description of the Great Work.

"Canalis" of Trenton contributes to the Empire a highly interesting article on the Murray Canal, which is rapidly approaching completion. Setting out with a historical sketch of the project from its inception in 1796 to its commencement in 1882, the correspondent gives to the late Jos. Keeler, M.P. for East Northumberland, and the Hon. Mr. Bowell, Minister of Customs, the credit which they deserve for securing the construction of this great commercial highway, and notes the letting of the contract to Messrs. Silcox & Mowry, and the turning of the first sod by Mrs. Keeler, widow of the energetic promoter, on Sept. 1st, 1882. The correspondent then, commencing at the eastern end, gives the following description of the work:

The cutting here is through a fine sand, like the ordinary beach sand, but further up the work I had heard that rock was met with. The banks are here "rip-rapped" with stone to protect them from the swell caused by vessels passing through and the spring floods. Proceeding westward along the bank I arrive at the first road bridge, of which there are three at different places, and a few rods farther on a railroad bridge, where the Central Ontario railway passes.

The road bridges, two of which are completed, swing on a huge central pier protected on either side by other piers, and are built by Mr. Robert Weddell, Jr., at this bridge works at Trenton; and bear out in their beauty and strength the great reputation which Mr. Weddell has earned by the quality of his work of being one of the foremost bridge builders in Canada. The railroad bridge, a fine structure, reflecting great credit on the builders, the Dominion Bridge Company of Hamilton, is also built of steel and swings in the same manner as the road bridges.

The banks of the canal on both sides of the bridges are protected by massive stone walls, laid up with cement. Further up the cutting is through a large marsh, called "Dead Creek Marsh," where it was supposed to be impossible to find solid bottom, and where the opponents of the present route prophesied the abandonment of the work on account of the quick-sands. But here, strange to relate, so far from being troubled with quick-sand, the banks are the best on the whole length of the cutting. There are three dredges at work here finishing the excavation and straightening the banks, thus completing the canal as far as navigation is concerned from the eastern extremity. Here also a great number of teams with scraper are busily engaged in leveling the banks in order to form a tow-path.

Passing the second road bridge a dam across the canal is arrived at, and another can be seen a short distance farther up. Between these dams, the water having been drawn off, appears the only rock met with in the whole route of about five miles. Had the contractors been aware of the existence of a bed of rock at this part of the cutting of such a size and of a quality equal to at least to any used on the works, they would have been saved the expense of procuring stone from the Point Ann quarries for the piers of the bridges. This rock is very irregular in its formation; rising to an elevation of about eight feet from the bottom on the southerly side, it gradually diminishes in thickness, in the form of steps towards the north, until it disappears altogether, or leaves only a thin ledge to be removed in order to obtain the required depth of the excavation. In this form the rock extends for about 2,800 feet, and was a complete surprise to the engineers, its existence being unknown until struck by the dredges. A large gang of men are at work at this point drilling and blasting out the rock to the required depth, and removing the pieces with derricks stations on each side of the canal, their long arms reaching half way from bank to bank. Another body of men are engaged in finishing the piers for the third and last road bridge, which is located here.

A short distance north from the third bridge site is a post office and quite a little village, where before the commencement of the canal there was only a solitary farm house. The buildings are principally boarding-houses put up for the accommodation of the workmen. The headquarters and office of the contractors being at this point of the works I proceed thither in quest of the foreman of the work, Mr. McAulicliff, who on being found, after laughing merrily at my muddy appearance, which was the result of my stepping upon an innocent looking clay bed and my sinking therein father and quicker that I expected, to the great merriment of the workmen, kindly consented to accompany me over the remainder of the southern bank, Mr. McAulicliff in the meantime giving me a detailed account of the chief point to be observed in the building of this canal.

Above this upper dam the canal is completed, the banks being all leveled, thus forming the tow-path, which looks like a wide and level roadway on each side. Par of this tow-path, or roadway, had to be made by filling in a large marsh with mud scows, making the operation very tedious. The banks, where formed of clay, which my companion informed me was the more unstable as compared with sand, are "rip-rapped" below the water line as in the eastern end.

Arriving at the end of the bank, we found two lines of piers extending out into Presqu'Isle harbour similar to those found at the entrance of the canal from the Bay of Quinte. I learned, on questioning my comrade, that these piers which are almost completely filled with stone, were be planked over with heavy plank, every other one being fitted with a tie post, so forming for all practical purposes, along dock on each side of the canal. A channel 200 feet wide and 16 feet deep has been dredged out in the harbour to the main channel leading out into the lake, which latter channel is 1,000 feet wide. From the appearance of it here, one can judge what the canal will be like when completed for the whole distance. The mouth of the canal on the western end finishes in a marsh, which the good old times when whisky brough a good price across the line, served as a hiding place for the small fleet of boats of the smugglers who lurked about it, on account of the willows which grow in profusion, and among which they could rest undetected through the day until night, when, under cover of its darkness, they could run across the lake, land their cargo in some secret and secluded place, and return the following night.

Standing on the most outermost end of the line of piers, one has a fine view of Presqu'Isle by or harbour and the surrounding shores. Across the bay can be seen the lighthouses and the white tents of the campers on Presqu'Isle Point, which is a noted summer resort for the surrounding townspeople. North-west about a mile from the north shore of the bay, among the trees, appears the picturesque little village of Brighton, with its pretty houses and wide clean streets.

I had now travelled the whole length of a canal having four and a half miles of clay cutting, which together with the formation of dredging and piers of convenient entrances at either end, make up an artificial channel six miles long, having a width on the bottom of eighty feet and two hundred feet at the ends, and which will have when entirely completed, at ordinary low water (or zero of the Toronto harbor gauges) a uniform depth of twelve feet six inches and sixteen feet in the channels leading to it. The actual uniform depth of water in the finished portion of the canal at present is thirteen feet six inches according to the Toronto harbour measurement. Towards the cost of construction of this work the Government, up to the 30th of June, 1887, the close of the fiscal year of 1886 and 1887, had paid a sum of $680,565.56.

Mr. J.D. Silcox, to whose energy the great progress towards and early completion of the canal is due, has superintended the operations of the contractors in person until a few weeks back, when he was compelled to relinquish his post through illness. In conversation with Mr. Mowry, one of the contractors, as the time of completion, I was told that the whole of the clay excavation and the trimming up of the banks would be completed this fall. Mr. Mowry said that he did not think it possible that all the work between the dams would be removed by that time; but since the winter would not hinder in the least the work of blasting it would be all removed by the spring, leaving then only the dams to be cut away to render the canal open to navigation. It might be necessary, he said, to keep a dredge at work nearly all next summer to dig down the slight irregularities which may be found in the bottom of the canal on sounding.

The advantages which are likely to arise from the construction of the work are then discussed in detail and finally summarized as follows.

I have now mentioned the several local advantages to be derived from the Murray canal, namely: The convenience to the grain trade in the vicinity, the stimulus given the manufacturers on the Bay of Quinte by extending the local and facilitating the foreign trades, the coasting and passenger traffic it would promote, and lastly the opening up of the mineral wealth of the community. These are, indeed, the chief apparent advantages, but any one for a moment supposing it would turn the course of the great highway of Canadian commerce on the great lakes in that direction is in error. Our commerce there is no longer carried on by means of canoes, but by vessels as large and strong as those with which our forefathers crossed the Atlantic and doubled Cape Horn. The cost of Prince Edward presents no longer any terror to the Canadian mariner. So it will appear that the only through trade which will pass through the canal will be that carried by passenger steamers from Toronto to Montreal, calling at intermediate ports. Of course, the canal will be important in very heavy gales, because then lake vessels having made Presqu'Isle harbor could pass through it and so pursue their voyage down the more placid bay, or come to anchor in a position which is perfectly secure to any number of ships that might find their way thither.

As a naval station, in case of war, it is pointed out, the canal would be of great advantage, and the opinions of General Sir James Carmichael Smyth and Col. McDougall to that effect are quoted.

The article is of such interest that we regret that lack of space prevents its reproduction in full.

APPENDIX C:
Has Shovel Used To Open Work On Murray Canal

Mrs. Archer Brown Proud of Silver-Plated "Shovel" Used By Her Grand-mother, Mrs. Joseph Keeler in 1882.

Most people call a spade a spade, but the good folk who arranged to have the first sod turned in the construction of the Murray Canal near Belleville in 1882, called the instrument that was used in the ceremony a "shovel" although it was only thirty-two inches long.

This information is contained in a letter belonging to Mrs. Archer Brown, Simcoe Street North, whose grandmother, Mrs. Joseph Keeler, actually broke the ground that started building operations on the canal that now joins the western end of the Bay of Quinte to Lake Ontario.

Mrs. Brown also has in her possession the spade that was used on that occasion. It is a silver-plated affair, with three small gold maple leaves fanning out from the wooden handle across the broad surface of the metal. The top end of the handle is shaped like a "T" and there is a gold-plated piece of filigree work at the top and another small piece of gold plate about half way down the handle, bearing the words:

"Presented to Mrs. Joseph Keeler by the citizens of Brighton and vicinity on the occasion of turning the first sod of the Murray Canal, August 31, 1883."

Mrs. Keeler was the wife of the member for Northumberland in the Canadian House of Commons who was greatly interested in shipping, and who was instrumental in having the canal built. For a number of years, he had a fleet of schooners that plied between Lakeport, near Brighton, and the American city of Oswego, across Lake Ontario.

The route was very stormy and treacherous, and many ships were wrecked. While he was in Parliament, Mr. Keeler pressed that a canal be built so that ships might pass into the Bay of Quinte and sail for many miles in the sheltered waters of the bay.

Did Not See Opening

That he was successful in his efforts may be seen in the building of the canal, but it was his misfortune never to witness the fruit of his labours. He died in January, 1881, at the age of fifty-nine.

The letter that was presented to Mrs. Keeler on the occasion of the sod turning reads as follows:

Mrs. Keeler: "Respected Madam: It is with a feeling of the most intense pleasure that we, the residents of this section of the country welcome you here today to perform so important and agreeable a ceremony. The work which occupied the time, attention and talents of your beloved husband and our late lamented and efficient representative in his parliamentary capacity finds its fitting inaugural in the events of this auspicious day. We are sure that while it must cause you, as it does us, keen sorrow to reflect that Mr. Keeler didn't live to see the results of his earnest efforts in behalf of the Murray Canal, yet a certain amount of consolation comes to us all when we have you here to take the initial step toward the fulfilment of such an important undertaking. You will therefore please accept this shovel with the inscription thereon, as a memento of the event, and turn the first sod of the Murray Canal. We trust you may, with all assembled here today, see the completion of this great work, and witness the meeting of the waters of Presque'Isle with the Bay of Quinte.

"(Signed) Thomas Webb, Chairman of the Celebration Committee. Brighton, 31st August, 1882,"

Note: A digital copy of this article was sent to me by Lenna Broatch on March 23, 2023.

APPENDIX D:
A Bit of First-Hand Information with Local Geography

The *Belleville Ontario-Intelligencer* runs an interesting column called "Looking Backward," which is of course, just what it says, i.e. items from their files of bygone years. On August sixteenth they reproduced a splendid bit of interesting and important news, exactly as they first sent it out in 1889, fifty years ago. …

50 Years Ago (August 16, 1889) … Through the courtesy of Mr. H. Corby, M. P., many citizens, including Mayor W. Jeffers Diamond and the members of the city council, Mr. Thomas Ritchie, President of the Board of Trade, Col. Strong, American consul here, and many other prominent citizens, joined a number of the prominent citizens of Trenton yesterday in formally opening the Murray Canal. Arriving at the canal three boats loaded with people; (with the Trenton Band playing on one of the boats) steamed through the canal as far as Brighton Wharf where a stop was made and where lay at the wharf the yacht *Surprise,* Commodore Forbes in command, which was the first boat to pass through the canal. Later the excursionists proceeded to Presqu'ile, which in the near future will become famous as a summer resort. At present there are many campers there. It was with regret that the party sailed for home. We noticed the following prominent persons present from a distance and from Trenton: Hon. Mackenzie Bowell, Minister of Customs; Mr. Cochrane, M.P.; Mr. H. Corby, M.P.; Mr. G. W. Ostrom, M.P.P.; Mr. D: Gilmour and Mayor Morrison of Trenton. Although there were four members of Parliament present in the party, no speeches were made."

Source

"Looking Backward: A Bit of First-Hand Information with Local Geography," *Belleville Intelligencer,* August 16, 1939, Brighton Digital Archives.

Note: This newspaper clipping was included in a batch of documents passed to Dan Buchanan by the Brighton Digital Archives (BDA). The BDA obtained a large number of documents from the Bangay family after the passing of Ralph Bangay. Some of those documents were provided to Dan Buchanan in order to facilitate the scanning and cataloguing process. While going through these documents, this item about the Murray Canal was discovered.

Thanks to the Bangay family for their generosity in donating these valuable pieces of our heritage landscape and thanks to the folks at the BDA for continuing their critical work of presenting our history to the community.

APPENDIX E:
The Naming of The Carrying Place Post Office

Up to the year 1913, there was no such thing as The Carrying Place Post Office. From the day it was opened (the records before Jan. 1, 1857 are not available), the post office was known as "Murray." Its first postmaster appears to have been Reuben Young, but no date was given. The record of postmasters is as follows.

October 1, 1885, T.J. Spafford

January 1, 1857, James L. B.ggar May ^ ^ ^ ^

February 1, 1894, Agnes F. Preston September 1, 1899, Harry A. Boyce March 17, 1911, C.M. Westfall

October 1, 1861, R.O. Dickens January 1, 1874, Peter Rowe February 1, 1877, R.J. Corrigan

Page 72
In 1911, when C.M. Westfall was appointed postmaster, he began an appeal for a change of name for the post office from "Murray," which was the name of the township in which half of the village was located, to "The Carrying Place" with its historical background. He solicited the help of Carrying Place-born W.H. Biggar, K.C., General Counsel for the Grand Trunk Railway Company of Canada.

The following letter is Mr. Westfall's copy of one sent by Mr. Biggar to Hon. L.P. Pelletier, Postmaster General, requesting the change of name:

March 7, 1913
Hon. L.P. Pelletier,

Postmaster General, Ottawa My Dear Minister:
I have just received a letter from the Postmaster of Murray Ont., stating that a petition is before you asking that the name of this post office be changed to "The Carrying Place." As I was born at "The Carrying Place" and both my Grandfather and Father for years filled the position as Postmaster, I naturally have considerable interest in the matter. I have often felt in my own mind that "Murray" was not an appropriate name for the post office, that being the name of the surrounding township in which several other post offices are located. I presume, however, that the post office at The Carrying Place was the first one established in the township. It certainly dates a long way back, as I have in my possession a letter addressed to my Grandfather then Postmaster which bears date 15th, February,

1825. You no doubt have been informed of the reasons why the name The Carrying Place was given to the isthmus which connects the Counties of Northumberland and Prince Edward and lies between the Bay of Quinte and what is known as Weller's Bay. The latter even within my recollection was landlocked with exception of a narrow channel, but long ago the beach between it and the Lake began to wash away and today the outlet is probably between one and two miles wide. I have always understood that in the early days it was much used as a portage for the reason that it not only formed a safer means of communication between the east and west by avoiding the dangers of navigation around the County of Prince Edward, but was also shorter. Be that as it may, the name The Carrying Place is that by which the way across this isthmus has for a very many years been known, and to my mind if you were to approve and authorize the change in the name of the post office, you would be only doing that which should have been done before. The Carrying Place has a distinctive place in the history of Upper Canada and a post office bearing the name could be immediately located by any student of history; while on the other hand to such a student the present name means nothing. Personally, I need not assure you that the position has my most cordial support and 1 feel certain that all the surviving members of our family would take the same view.

Yours faithfully, (Sgd.) W.H. Biggar
Page 73

Apparently, Mr. Westfall's petition and Mr. Biggar's personal appeal impressed the Postmaster General sufficiently, for "The Carrying Place Post Office became a reality.

Source: Seventh Town Historical Society, "Chapter 6: The Carrying Place," in *7th Town/Ameliasburgh Township, Past & Present*, (Milton: Global Heritage Press, 1999), 71 & 73.

APPENDIX F:
Land Expropriation Records

The information in this list represents the pieces of land expropriated for the Murray Canal. The records are available in the Ontario Land Registry Records, which the author has copied from OnLand.ca. The details for these transactions are also available in the records for the people involved in www.treesbydan.com. In all of these transactions, the grantee is "Her Majesty the Queen."

Murray Township, Carrying Place Lots

Grant #3167 shows that the amount of 2 and 776/1000 acres was purchased from Lewis Latour and wife for $175, from Carrying Place Lot 13, June 18, 1883.

Grant #3317 shows that the amount of 29/100 of an acres was purchased from David John Huffman and wife for $1, from Carrying Place Lot 13, November 3, 1883.

Grant #3388 shows that part of Carrying Place Lots 12, 11, and 10 were puchased from Benjamin Rowe and wife for $2,500 (total for all), June 22, 1883.

Grant #3172 shows that 9 and 216/1000 acres from Carrying Place Lot 9 was purchased from the Trustees of Murray School District 1 for $500, May 31, 1883. (Lot 9 is shown as "School Land" in the 1878 Belden County Atlas.)

Grant #3178 shows that part of the north-west part of Carrying Place Lots 8 and 7 were purchased from Benjamin Rowe and wife for $1, June 22, 1883.

Grant #3120 shows that 92/100 acres of Carrying Place Lot 6 was purchased from Thomas A. Porter and wife for $75, April 21, 1883. (Porter was a son-in-law of Ben Rowe.)

Murray Township, Concession C

Grant #3181 shows that 2 and 757/1000 acres from the north-west part of Lot 8, and the north part of Lot 9, Concession C, Murray Township was purchased from Stephen H. Flindall and wife for $120, August 11, 1883.

Grant #3186 shows that 134/1000 acres from the north of Lot 8, Concession C, Murray Township was purchased from George J. Flindall and wife for $60, August 15, 1883.

Decree #556 shows that 45/100 acres from Lot 10, Concession C, Murray Township was granted to Her Majesty the Queen by the Court of Cancery, October 8, 1884.

Decree #540 shows that 2 and 58/100 acres from Lot 11, Concession C, Murray Township was granted to Her Majesty the Queen by the Court of Cancery, February 14, 1884.

Grant #3085 shows that part of the north-west half of Lot 13, Concession C, Murray Township was purchased from Peter and Hannah Gould for $10, February 17, 1883.

Grant #3089 shows that all of his interest in part of Lot 13, Concession C, Murray Township was purchased from Henry S. Allard for $10, February 26, 1883.

Grant #3110 shows that 1 and 62/100 acres in the north part of Lot 14, Concession C, Murray Township was purchased from Peter Gould and wife for $800, February 17, 1883.

Grant #3082 shows that 3 and 13/100 acres of the north part of Lot 15, Concession C, Murray Township was purchased from William H. Goldsmith and wife $250, February 17, 1883. (William H. Goldsmith was husband of Sarah Ann Gould, daughter of Peter Gould.)

Grant #3052 shows that 4 and 82/100 acres of the north part of Lot 16, Concession C, Murray Township was purchased from Cadwell K. Stoneburgh and wife for $400, January 18, 1883.

Grant #3041 shows that 3 and 32/100 acres of the north-west quarter of Lot 17, Concession C, Murray Township was purchased from Esther A. Lee and husband for $125, January 4, 1883.

Grant #3053 shows that 3 and 32/100 acres of the north-west quarter of Lot 17, Concession C, Murray Township was purchased from Mary Goldsmith and husband for $75, January 18, 1883.

Grant #3090 shows that 2 and 95/100 acres of the north part of the east half of Lot 17, Concession C, Murray Township was purchased from Philip H. Lawson and wife for $250, February 17, 1883.

Grant #3040 shows that 7 and 47/100 acres of the north half of Lot 18, Concession C, Murray Township was purchased from William Lovett and wife for $500, January 4, 1883.

Grant #3081 shows that 7 and 63/100 acres of the north part Lot 19, Concession C, Murray Township was purchased from Charles Lee and wife for $600, December 29, 1882.

Grant #3030 shows that 24/100 acres of the north-east quarter of Lot 20, Concession C, Murray Township was purchased from Samuel May and wife for $20, December 29, 1882.

Grant #3032 shows that 1 and 99/100 acres of the north quarter of Lot 20, Concession C, Murray Township was purchased from Joseph Wilson and wife for $150, December 29, 1882.

Grant #3038 shows that 6 and 13/100 acres of the north half of Lot 21, Concession C, Murray Township was purchased from Jonathan Hutchinson and wife for $3,000 (total for all) January 4, 1883.

Grant #3039 shows that 8 and 54/100 acres of the north half of Lot 21, Concession C, Murray Township was purchased from William Lovett and wife for $2,500 (total for all) January 4, 1883.

Grant #3031 shows that 7 and 58/100 acres of the north half of Lot 22, Concession C, Murray Township was purchased from Charles Clindinin and wife for $1,000, December 29, 1882.

Murray Township, Concession B

Grant #3397 shows that 545/1000 acres of the south part of the west half of Lot 13, Concession B, Murray Township was purchased from Sylvester Sills and wife for $40, May 31, 1883.

Grant #3146 shows that 5 and 844/1000 acres of the south part of Lot 14, Concession B, Murray Township was purchased from John May and wife for $385, May 31, 1883.

Grant #3145 shows that 2 and 65/100 acres in the south-east part of the south part of Lot 15, Concession B, Murray Township was purchased from George H. May and wife for $180, May

31, 1883.

Decree #528 shows that 6 and 838/1000 acres in the south-west part of the south part of Lot 15, Concession B, Murray Township was granted to Her Majesty the Queen by the Court of Chancery, September 11, 1883.

Decree #530 shows that 2 and 65/100 acres in the south-east part of the south part of Lot 15, Concession B, Murray Township was granted to Her Majesty the Queen by the Court of Chancery, October 12, 1883.

Grant #3115 shows that 4 and 99/100 acres in the west half of the south part of Lot 16, Concession B, Murray Township was purchased from Samuel F. May and wife for $210, March, 31, 1883.

Grant #3143 shows that 7 and 29/100 acres in the east half of the south part of Lot 16, Concession B, Murray Township was purchased from William Evans and wife for $420, May 31, 1883.

Grant #3093 shows that 84/100 acres in the east half of the west half of the south part of Lot 17, Concession B, Murray Township was purchased from Thomas P. Powers and wife and Eliza Powers, widow of James Powers, deceased for $75, February 17, 1883.

Grant #3088 shows that 0.54 of an acre in the south-east part Lot 18, Concession B, Murray Township was purchased from Joseph T. Pelkey and wife for $50, February 26, 1883.

Brighton Township, Concession C

Grant #3494 shows that 20 acres in the north part of Lots 23, 24 & 25, Concession C, Brighton Township was purchased from John McMaster and wife for $1,000 (in total), December 29, 1882.

Grant #3569 shows that 97/100 of an acre in the north-west corner of Lot 26, Concession C, Brighton Township was purchased from Hiram G. Lawson and wife for $10, May 31, 1883.

Grant #3545 shows that 1 and 644/1000 of an acre in the north part of Lot 27, Concession C, Brighton Township was purchased from Aaron W. Talmage and wife for $50, March 31, 1883.

Grant #3556 shows that 0.556 of an acre in the north part of Lot 28, Concession C, Brighton Township was purchased from Martha and Daniel Church for $40, March 21, 1883.

APPENDIX G:
Supporting Material

Readers who are interested in seeing more details can go to my web site:

www.danbuchananhistoryguy.com

There you will find downloadable PDF files containing:
- a list of illustrations;
- a much more detailed Notes and References document;
- a sources list;
- a searchable version of the Powles Report; and
- a compendium of other source documents.

APPENDIX H:
Note from Colin Powles

In August 2023, the author of this book made contact with Mr. Colin Powles, the author of the Powles Report. Mr. Powles was generous with his time and agreed to provide a summary of his involvement with the creation of the document.

Here is his summary:

The report originated as a summer co-op project with the Ontario Regional Office of Parks Canada, in Cornwall, Ontario. I was completing a Master's degree in Public History at the University of Western Ontario, and had previously worked with Parks Canada to write the history of the Kirkfield Lift Lock on the Trent Severn Waterway (TSW) in 1989. For the Murray Canal project, I was hired on a 4-month summer contract in 1991 to research and write a construction and maintenance history of the canal. This was to be used by Parks interpretive staff and for informational plaques. I worked under the supervision of Parks Canada historian Dr. Paul Couture, who was then the official historian of the Rideau Canal as well.

Working out of Cornwall in the pre-internet days, most of the research was done at the National Archives in Ottawa, using extensive resources primarily from the Department of Railways and Canals record group. This entailed several weeks in Ottawa, but as the tortuous story emerged of the canal's many surveys and delays, other trips were needed for a more local perspective. I travelled to archives in Belleville, Picton and Kingston, and to the TSW offices in Peterborough to try and unearth some of the reasons that the work was delayed for so long. For information on daily life working on the canal, I arranged oral history interviews with several canal employees and bridge masters who still lived in the Carrying Place area.

Once the research was completed, I wrote up the report and submitted it to Dr. Couture for editing and revisions. The manuscript was completed and approved as a Parks Canada report in late August of 1991 and I returned to London in September to complete my MA course work before graduating in April of 1992 from the University of Western Ontario.

Colin Powles
August 2023

BIBLIOGRAPHY

Andrews, Margaret, et al. *Pictorial Brighton 1859–1984*. Brighton: 125th Anniversary Book Committee, 1984.

Calnan, Marion Mikel, Peggy Dymond Leavey, and Julia Rowe Sager. *Gunshot and Gleanings of the Historic Carrying Place, Bay of Quinte*. Bloomfield: 7th Town/Ameliasburgh Historical Society, 1987.

Correspondence of Lieut. Governor John Graves Simcoe, The. With Allied Documents Relating to His Administration of the Government of Upper Canada, Brigadier General E.A. Cruikshank, LL.D., F.R.S.C., collected and ed., for the Ontario Historical Society. *Volume V. 1792–1796 (Supplementary)*. Toronto Historical Society, 1931. PDF copy downloaded from Internet Archives.

French, Orland, ed. *Heritage Atlas of Hastings County*. Belleville: Wallbridge House Publishing, 2006.

Jarrett, Thomas. *The Evolution of Trenton, Ontario: A Railway, Power and Manufacturing Centre*. Trenton: Publisher not identified, 1914. https://www.canadiana.ca/view/oocihm.80996/1.

Muntz, Madelein "Peggy." *John Laing Weller C.E., M.E.I.C.: "The Man Who Gets Things Done."* St. Catharines: Vanwell Publishing Limited, 2007.

Plomer, James, and Alan R. Capon. *Desperate Venture: The Central Ontario Railway*. Belleville: Mika Publishing Company, 1979.

Powles, Colin. "A Construction, Operations and Maintenance History of the Murray Canal." Unpublished manuscript for Canadian Parks Service, Ontario Regional Office, Summer 1991, typescript. http://danbuchananhistoryguy.com/uploads/1/1/5/0/115043459/murray_canal_colin_powles_report_1991_searchable.pdf.

Seguin, Marc. *For Want of a Lighthouse: Building the Lighthouses of Eastern Lake Ontario 1828–1914*. Bloomington: Trafford Publishing, 2015.

Simcoe, Elizabeth, and J. Ross Robertson. *The Diary of Mrs. John Graves Simcoe*. Toronto: William Briggs, 1911.

The Municipality of Brighton. *That's Just the Way We Were. Brighton Memories*. Brighton: Brighton History Book Committee, 2006.

Tobey, William M. *The Tobey Book*, Barbara Nyland, ed. Unpublished manuscript, July 1975, typescript. http://danbuchananhistoryguy.com/uploads/1/1/5/0/115043459/the_tobey_book.pdf.

Online Resources

Ancestry.ca, https://www.ancestry.ca/search/

Brighton Digital Archives, https://vitacollections.ca/brightonarchives/search

Broatch, Lenna, Historian of Cramahe Township

Canada Gazette, Government of Canada

Canada Historic Sites, https://parks.canada.ca/lhn-nhs

Canadian County Atlas Digital Project, https://digital.library.mcgill.ca/countyatlas/search.htm.

Canadian Encyclopedia, https://www.thecanadianencyclopedia.ca/en/

Canal River Trust, https://canalrivertrust.org.uk/enjoy-the-waterways/canal-history/the-first-canal-age-canal-history

Chronology of Kingston Architecture, Jennifer McKendry, http://www.mckendry.net/CHRONOLOGY/chronology.htm

Community Archives of Belleville and Hastings County, https://discover.cabhc.ca/

Dan Buchanan, The History Guy of Brighton, http://danbuchananhistoryguy.com/index.html

David Rumsey Map Collection, The, https://www.davidrumsey.com/

Dictionary of Canadian Biography, http://biographi.ca/en/

Government of Canada, Lachine Canal National Historic Site, https://parks.canada.ca/lhn-nhs/qc/canallachine/culture/histoire-history/histoire-history

Hastings County Historical Society, https://hastingshistory.ca/

HeinOnline, Law Journal Library, https://home.heinonline.org/content/Law-Journal-Library/

Heritage Cramahe, https://heritagecramahe.ca/

History Lives Here, Peter Lockyer, https://historyliveshere.ca/

Kingston Chronicle & Gazette, Digital Kingston.

Library and Archives Canada, https://digital.library.mcgill.ca/countyatlas/search.htm.

Murray Canal District Organization, https://murraycanaldistrict.ca/

Ontario Land Registry Records, https://www.onland.ca/ui/

Parks Canada, https://parks.canada.ca/index

St. Lawrencepiks – Seaway History, http://stlawrencepiks.com/seawayhistory/

TreesByDan, http://www.treesbydan.com/main.htm

Wikimili, https://wikimili.com/en/

Wikipedia, https://www.wikipedia.org/

WikiTree, https://www.wikitree.com/wiki

ILLUSTRATION CREDITS

Front Cover
Murray Canal, photo by Author, 2022.

Inside Front
Murray Canal, aerial photo by Sean Scally, 2022.
Murray Canal Looking East, aerial photo taken Sean Scally.

Dedication
Florence Chatten, photo by author, 2018.

Preface
Three Books, composite of images of front covers, by author.
The Carrying Place, front page of presentation by author, Oct 18, 2022.
The Murray Canal, front page of presentation by author, February 23, 2023.
Signature of Dan Buchanan, by author
The History Guy, taken by Ken Strauss, Cobourg & District Historical Society, Nov 18, 2018.

Powles Rep
The Powles Report, snip of font page, by author.

Chapter 1
Lieutenant Governor John Graves Simcoe, BIO, *Dict of Canadian Biography*.
The Carrying Place, sketch from Ed Burtt Collection.
Upper Canada 1800, David Rumsey Map Collections, illustrations by author.
Canal Reserve, Digital County Atlas Project, segment of Murray Twp.
Asa Weller's Batteau Portage Service, painting by Bowen P. Squire, digital image from Sean Scally.
Colonel Richard Bullock, United Lodge No. 29, Brighton, history by Edfar W. Pickford, Brighton Digital Archives.
Timber Raft, from "Ironmasters of Upper Canada," permission of Sean Scally.
Henry Ruttan, BIO, *Dictionary of Canadian Biography*.

Chapter 2
Colonel By, Rideau Canal, BIO, *Dictionary of Canadian Biography*.
Sir John Colborne, BIO, *Dictionary of Canadian Biography*.
The Carrying Place – Governor General Simcoe, Kingston Chronicle & Gazette, Google Newspapers.
Nicol Hugh Baird, BIO, *Dictionary of Canadian Biography*.
Rebels March Down Yonge Street to attack Toronto, 1837, C. W. Jefferys, in *Canada 1812–1871: The Formative Years*.

Lord Durham, The Canadian Encyclopedia.
Charles Poulett Thomson, Archives of Ontario.
Board of Works Statute, *The Canada Gazette*, Library and Archives Canada.
Murray Canal Surveys to 1845, Wikipedia, illustrated by author.

Chapter 3

Railway Construction Gang, *The Canadian Encyclopedia*.
Gilmour Lumber Company, Wikipedia.
Railway Locomotive, The Canadian Encyclopedia.
Wellers Bay vs. Presqu'ile Bay, segment of Anderson's 1893 Survey, Library and Archives Canada.

Chapter 4

James L. Biggar, M.P., Wikipedia.
Canadian Parliament Buildings, Library and Archives Canada.
Rowan Survey 3 Routes, Wikipedia, illustrated by author.
Presqu'ile Bay and Middle Ground Shoal, segment of Anderson Survey 1893, Library and Archives Canada, illustrated by author.
John A. Macdonald, Library and Archives Canada.
Joseph Keeler III, Heritage Cramahe.
James L. Biggar, Wikipedia.
The Canals of Canada, John P. Heisler, Department of Indian Affairs and Northern Development, Ottawa, 1973, *Canada Historic Sites: Occasional Papers in Archaeology and History*, No. 8, p. 3.
Map of Trent Navigation and the Murray Canal, The Canals of Canada, p. 33 of 174.

Chapter 5

Marine Disaster 1879, Marc Seguin, *For Want of a Lighthouse*, p. 272.
Mackenzie Bowell, The Canadian Encyclopedia.
Sir Charles Tupper, BIO, *Dictionary of Canadian Biography*.
Joseph Keeler III, M.P., 1868, Topley, William, Library and Archives Canada.
Appropriations for the Murray Canal 1881, Order in Council No. 1881-0555, Library and Archives Canada, RG2, Privy Council Office, Series A-1-a, Vo. 402, reel C-334.
Adoption of Route, Order in Council No. 1882-1081, Library and Archives Canada, RG2, Privy Council Office, Series A-1-a, Volume 415, reel C-3338.
James Simeon McCuaig, Parliament of Canada, House of Commons.
John Milton Platt, FindaGrave, Memorial ID: 1042254490.

Chapter 6

Murray Canal Full Length, Google Maps segment, illustrated by author.
Bridge Piers Off-Set for Lumber Barges, photo by author, 2022.
Joseph Keeler III, Heritage Cramahe.
Octavia Phillips Keeler, Cramahe Public Library, Cramahe Digital Archives.
Silver Spade Used to Turn Sod for Murray Canal, photo by Ron Waddling 2013.
Mackenzie Bowell, The Canadian Encyclopedia.
Wheeled Earth Scraper, Ameilaisburgh Museum, photo provided by Sean Scally.

Chapter 7
Bridges Over the Murray Canal, Canadian County Atlas Project, segment of Murray Canal map, illustrated by author.
John Laing Weller, Madelein "Peggy" Muntz, *John Laing Weller, C.E., M.E.I.C.: "The Man Who Gets Things Done,"* p. 11.
Gould Clearing, Canadian Digital County Atlas Project, segment of Murray Twp., illustrated by author.
Murray Canal Construction — Dam, William M. Tobey, *The Tobey Book*, p. 483.
Murray Canal Construction — Sand Banks, William M. Tobey, *The Tobey Book*, p. 483.
Murray Canal Construction — Dredge, William M. Tobey, *The Tobey Book*, page 483.
Murray Canal Construction — Swing Bridge Pier, image provided by Sean Scally.
R. Weddell Dredge, Thomas Jarrett, *Evolution of Trenton*, p. 38, Canadiana.ca.
R. Weddell Dredge, Thomas Jarrett, *Evolution of Trenton*, p. 38, Canadiana.ca.
R. Weddell Dredge, Thomas Jarrett, *Evolution of Trenton*, p. 38, Canadiana.ca.
R. Weddell Dredge, Thomas Jarrett, Evolution of Trenton, p. 38, Canadiana.ca.
Railway Bridge Diversion, Canadian Digital County Atlas Project, segment of Murray Township, illustrated by author.
Bridges Built by Trenton Bridge and Engine Works, Flickr.
R. Weddell Dredge c. 1910, Marion Mikel Calnan, Peggy Dymond Leavey, and Julia Rowe Sager, *Gunshot and Gleanings of the Historic Carrying Place*, p. 91.

Chapter 8
Murray Canal Demonstration Poster, image donated by Roy Bruce, Belleville Community Archives.
Bay of Quinte Bridge, Fano, Donna, article in Outlook newsletter, Community Archives of Belleville and Hastings County.

Chapter 9
The Murray Canal, Canalis, *Belleville Daily Intelligencer*, August 23, 1888, Community Archives of Belleville and Hastings County.
Murray Canal Sign, photo by author.
Lovett on the Map, Google Maps, segent of Lovett and Murray Canal.

Chapter 10
Fixing the Rates, Order In Council No. 1890-1216, Library and Archives Canada, RG2, Privy Council Office, Series A-1-a, Vol. 558, reel C-3407.

Chapter 11
Lock Tender Turning Big Handle, image by permission of Sean Scally, from his video "The Carrying Place: Kente Portage."
Sir John A. Macdonald 1890, Library and Archives Canada.
Bay of Quinte Bridge, Community Archives of Belleville and Hastings County.
Richelieu & Ontario Navigation Co., Picturesque Bay of Quinte, Community Archives of Belleville and Hastings County.
Steam Ship Varuna, Picturesque Bay of Quinte, Community Archives of Belleville and Hastings County.

Steam Ship Brockville, Community Archives of Belleville and Hastings County.

Chapter 12
Pier Light, Marc Seguin, *For Want of a Lighthouse*, p. 324.
Electric Railway Proposal Sketch, Order in Council No. 1899-0535, Library and Archives Canada, RG2, Privy Council Office, Series A-1-a, Vol. 777, reel C-3770. Sketch illustrated by author.
Apple Pickers, Margaret Andrews et al., *Pictorial Brighton 1859–1984*, p. 62.

Chapter 13
Smithfield Bridge Collapse, The Municipality of Brighton, *That's Just the Way We Were, Brighton Memories*, p. 147.
Smithfield Bridge Piers Today, photo by author, 2022.
Carrying Place Road Bridge, photo by author, 2022.

Chapter 14
Map of Land EXpropriations by Area, Canadian Digital County Atlas Project, segment of Murray Township map, illustrated by author.
Map of Carrying Place Lots, Canadian Digital County Atlas Project, segment of Murray Township map, illustrated by author.
Map of Murray Township, Concession C, Canadian Digital County Atlas Project, segment of Murray Township map, illustrated by author.
Map of Murray Township, Concession B, Canadian Digital County Atlas Project, segment of Murray Township map, illustrated by author.
Map of Brighton Township, Concession C, Canadian Digital County Atlas Project, segment of Murray Township map, illustrated by author.
Lovett Canal House, image linked to family tree of janice1661 on Ancestry.ca, originally shared by Mama Sinclair, Sept. 24, 2011.

Chapter 15
Quinte West Created January 1, 1998, Orland French, ed., *Heritage Atlas of Hastings County*, p. 200, by permission of Hastings County.
Tug Boat Holds Murray Canal Display, photo by author, 2022.
Millennium Trail Sign, photo by author, 2022.
Dan's Bike, photo by author, 2014.
Bridge at County Road 64, photo by author, 2014.
Murray Canal Celebrates 125th Birthday, History Lives Here, https://historyliveshere.ca/.
Poster for Meeting re Single Lane Bridge, Murray Canal District Organization.
High Water at old Bridge, photo by author 2017.
Preparation for Bridge Replacement, photo by author, 2018.
Working on the New Bridge, photo by author, 2018.
New Bridge at County Road 64, photo by author, 2018.
Full Two Lanes With Pedestrian Walkway, photo by author, 2022.

INDEX

A

Adolphus Reach, 1, 52
Allard, Enoch, 63
Allard, Hannah Maria (wife of Peter Gould), 63
Allard, Henry S., 63
Alyea, Allan, 47, 60, 68
Alyea, Fred (father of Allan), 68
Alyea, James Henry, 68
Amalgamation, 70
Ameliasburgh Museum, 27
Ameliasburgh Township, Prince Edward County, 13, 62
American Civil War, 10, 11
Anderson, William, 51
Athol Township, 22
Austin, G. E., 19, 20
Austin, Natalie (Trent-Severn Waterway), 73

B

Baird, Nicol Hugh, 6, 7
Baltimore Development Services, Cobourg, 59
Batteau, 1, 3, 56
Bay of Quinte Bridge, 37, 49
Begg, Alex, 9
Belden County Atlas, 3, 62
Belleville, 18, 19, 26, 31, 37, 47, 48, 49, 50, 53, 54, 62
Belleville City Council, 9, 18
Belleville Daily Intelligencer, 33, 38
Belleville Portland Cement Company, Point Anne, 52, 54
Berkshire County, Massachusetts, 15
Biggar, Charles, 12
Biggar, James L., 12, 15, 16, 18, 22, 58
Biggar, Mary, 12
Biggar, William Hodgins, 58
Blairton Mine, The, 19
Board of Works, 6, 7, 9

Bonter, Catherine Maria, 62
Bonter Sidewalk, The, 62
Bowell, Hon. Mackenzie, 19, 26, 32, 35, 43, 45, 52
Boyer, Ruth (granddaughter of Joseph Keeler III), 26
bridgetender, 42, 47, 55, 56, 58, 59, 60, 61, 67, 68, 69, 72
Brighton, 4, 8, 9, 25, 26, 29, 35, 40, 41, 44, 50, 52, 54, 56
Brighton and Seymour Gravel Road, 8
Brighton Road, 24, 25, 30, 38, 39, 59, 64, 66, 68
Brighton Road bridge, 24, 28, 36, 40, 42, 54, 55, 57, 58, 59, 68, 72
Brighton Road Swing Bridge, 73, 75
Brighton Sentinel, 9
Brighton Township, Northumberland County, 8, 9, 25
Brighton Wharf, 43
British Empire, 15
British North American Colonies, 6, 11
Butler's Rangers, 63
By, Colonel, 5, 6
Bytown (Ottawa), 5

C

Campbellford, 8
Canada, 6, 7, 12, 15, 16, 20, 25, 35, 36, 39, 43, 44, 45, 48, 49, 53, 58, 63, 74
Canada, Dominion of, 15, 45
Canada, Lower, 6
Canada, Province of, 6, 12, 14
Canada Cement, 53, 54
Canada East, 6
Canada Gazette, 49
Canada, Upper, 6, 7, 8, 9, 12, 15, 20
Canada West, 6
Canadian National Railway, 55
Canadian Parks Service, xiv
Canadians, 11
Canal Reserve, 2, 3, 4, 30, 62, 64, 68, 69
Canalis, 38, 39, 40, 41

Canals, Golden Age of, 1
Cape Horn, 41
Cape Vincent, New York, 18
Carrying Place, 12, 22, 56, 62, 67, 70
Carrying Place, The, 1, 3, 4, 12, 56, 58, 62
Carrying Place bridge, 59
Carrying Place lots, 61, 62
Carrying Place Road, 58
Cataraqui, 1
Central Bridge Company, Trenton, 57, 58
Central Ontario Railway, 23, 28, 34, 36, 48, 51, 53, 55, 71
Chapel Street, Brighton, 71
Chase, John, 62
Chatten, Ethel (wife of Gordon John Harvey), 68
Church, Daniel, 67
Church, Martha (wife of Daniel Church), 67
Clarendon Hotel, Brighton, 25
Clarke House, 26
Clindinin, Charles, 64
Clindinin, James Nelson, 64
Cobourg, 4, 12, 15, 19, 29, 43, 59, 66
Cochrane, Edward, 33
Coe Hill, Hastings County, 19, 48
Colborne, 15, 22
Colborne, Sir John, 5
Collingwood, 44
confederation, 11, 12, 14, 15
Conservative Party, 16, 19, 22, 33
Conservatives, 47
copper pier lamps, 46
Corby, Henry, 37, 43
Cornwall, 6, 17, 20
County Road 64 (Brighton Road), 40, 71, 72, 73, 74, 75
Court of Chancery, 63
Cramahe Township, Northumberland County, 2, 63
Crown Land, 4, 30, 63, 64
Cuthbert, Alexander, 43

D

Daily Intelligencer, 16, 33, 38, 49
Danforth Road, 4, 15, 56

Davis, Martha (wife of Richard Hanna), 67
Dead Creek, 3, 4, 13, 21, 28
Dead Creek Marsh, 24, 29, 30, 39, 63
Desperate Venture, 28, 60
Diamond, Mayor W. Jeffers, 43
Dominion Bridge Company, 33, 36
Dominion Land Surveyor, 25
dredge, 13, 18, 22, 23, 28, 29, 30, 31, 32, 33, 39, 40, 49, 51, 54
Durham, Lord, 6, 7
Durham Report, 6

E

earth scraper, 26, 27, 29
Edinburgh, Scotland, 33
electric lighting, 59, 69
electric railway, 51, 52, 53
electricity, 59, 60
English Settlement Road, 4, 67
Erie Canal, 15

F

Family Compact, 5, 6
Fenian raids, 11
Fitzgerald, Fred, 46
Flindall, George James, 63, 68
Flindall, John Morris, 63
Flindall, Stephen Henry, 63
Flint and Holden Mills, Belleville, 47, 52
Forbes, Commadore Alexander, Jr., 43
Forbes, Alexander, Sr., 43
Francis, Charles, 43

G

Gananoque, 42
Gear, Mr., 43, 44
General Surveyor's Office, 2
Gildersleeve (Shipping Line), 50
Gilmour, David, 43
Gilmour Lumber Company, 9, 47, 48, 52
Glengarry County, 63
Goldsmith, Gilbert, 64
Goldsmith, William Henry, 64
Goodrich, Wesley, 68, 69

Google Maps, 40
Gosport, 52
Gould Clearing, 29, 30, 63
Gould, John, 63
Gould, Joseph, 30
Gould, Lydia (wife of Abraham Stoneburgh), 64
Gould, Mary Jane (wife of Joseph Wilson), 64
Gould, Nathan (brother of John), 63
Gould, Peter, 63, 64
Gould, Sarah (wife of William H. Goldsmith), 64
Gould, Seth Burr (brother of John), 63
Government House, Ottawa, 45
Grafton, 67
Grand Trunk Railway, 8, 9, 19. 58
Green, Ernest, 55

H

Hamilton, 50
Hamilton, Robert, 12
Hanna, George, 67
Hanna, Harold Grant, 67, 69
Hanna, Herb, 67
Hanna, Hugh Burton (son of John and Susan), 67
Hanna, John, 67
Hanna, Richard, 67
harbor of refuge, 7, 10, 14
Harbour Street, Brighton, 35
Harries, C. A., 46
Harvey, Gordon John, 68
Harvey, John Robert (Jack), 67, 68
Hastings County, 2, 19, 48, 49, 70
Hastings County Council, 10
Heisler, John P., 16
Hetherington, Lucy (wife of Samuel May), 65
Highway 33 (Trenton Road), 31, 57, 70, 71, 72
Hill, Mr., 57
Hilton, 8
History Guy of Brighton, 71, 72
Horsley, William Thomas "Willie", 68, 69
Huffman, David John, 62
Hutchinson, Jonathan, 64
Hutchinson Road, 56

I

InQuinte News, 73
Indian Affairs and Northern Development, Department of, 16

J

Johnson, William H., 67, 69

K

Keefer, Samuel, 9, 10
Keeler, Joseph III, 12, 15, 16, 19, 20, 22, 25, 26, 42, 66
Keeler, Joseph "Old Joe", 15
Keller, Joseph A. (son of Joseph III), 66
Keeler, Thomas Phillips, 42, 43, 52, 68
Kingston, 1, 5, 14, 16, 29, 50, 56
Kingston Board of Trade, 14
Kingston Chronicle and Gazette, 5
Kingston Navigation Company, 50

L

Lachine Canal, 5, 17
Lake Erie, 17
Lake Ontario, 1, 3, 4, 5, 8, 9, 10, 13, 14, 15, 16, 21, 23, 40, 44, 49, 52, 53, 60, 74
Lakeport, 15
land, expropriation of, 24, 25, 27, 61, 62, 64, 65, 67
land valuators, 24, 66
Laroche, E. H., 36
Lawson, Hiram G., 67
Lawson, Katie, 64
Lawson, Phillip H., 64
Lawson Settlement, 64
Lee, Charles, 64
Lehigh Portland Cement Company, Point Anne, 52, 53, 54
Library and Archives Canada, 21, 44, 61
lightkeeper, 46, 47
Lockyer, Peter, 72
Lovett, 40, 70
Lovett Bridge Canal House, 68
Lovett, Allen Edwin "Marcus", 68
Lovett, John, 25

Lovett, William, 25, 30, 39, 40, 64, 68
Lyons, Mr., 7

M

Macauly, Mr., 4
Macdonald, Sir John A., 14, 15, 16, 18, 22, 33, 35, 42, 44, 45, 48, 49, 71, 72
Macdonald Contracting Company, Toronto, 54
Macdonald Heritage Trail, 72
Macdonald Project of Prince Edward County, 72
Mackenzie Rebellion, 6
Magnet, 49
Major N. H. Ferry, 54
Manley, Mary M. (wife of John McMaster), 66
Marine and Fisheries, Department of, 46, 51
Marmora, 19, 34, 48, 59
Masonic Lodge, Trenton, 33
May, George H. (son of Samuel & Lucy), 65
May, John (son of Samuel & Lucy), 65
May, John (father of Samuel Fletcher), 65
May, Samuel, 64
May, Samuel Fletcher, 65
McAulicliff, Mr., 38, 40
McAuliff, M., 46
McCadden, J., 42
McCuaig, James Simeon, 22
McKenzie, Mary (wife of Gilbert Goldsmith), 64
McMaster, John, Jr., 66, 67
McMasters, George, 66
McMasters, Samuel, 66
McPhail, Susan (wife of John Hanna), 67
Meyers, John Walden, 62
Meyers, Mary (wife of John Row), 62
Middle Ground Shoal, 10, 13, 14, 23, 28, 30, 32, 49
Millennium Trail, 71
Molson Bank, Trenton, 43
Montreal, Quebec, 4, 5, 8, 9, 17, 33, 41, 50, 53, 58
Montreal Board of Trade, 10
Morrison, Mayor, 43
Mowry, H. J., 23, 26, 28, 29, 38, 41, 44
Murray Canal Demonstration, 35
Murray Canal District Organization, 73
Murray Township, Northumberland County, 2, 3, 4, 8, 12, 13, 15, 25, 61, 62, 63, 64, 65, 67, 70
Mutual Steamship Company, 53

N

New Hampshire, 63
New York State, 8, 9, 12, 15, 16, 44, 48, 63
Newark, Upper Canada, 1
Newcastle, Upper Canada, 2
Newcastle District, 4
Niagara District, 12, 63
Northumberland East, Riding of, 12, 15, 33
Northumberland County, 4, 13, 70

O

Old Percy Road, 8
Onandaga, 1
Ontario Department of Highways, 57
Ontario Department of Public Works, 9, 13, 19, 53
Ontario Land Registry Records, 12, 61
Ontario Northern Railway, 55
Order in Council, 45, 49, 51, 64, 66
Ostrom, G. W., 43
Oswego, New York, 15, 18, 48
Ottawa (Bytown), 5, 12, 17, 18, 20, 42, 43, 45, 48, 58

P

Page, John, 13, 14, 33
Parks Canada, 73, 74, 75
Parks Canada Infrastructure Project, 73
Pelkey, Joseph Thomas, 65
Perley, H. T., 19, 20
Peterborough, 24, 59
Phillips, Octavia (wife of Joseph Keeler III), 25, 26
Phillpotts, Lt. Col. E., 7
Phin, W. E., 54
Phinn, Cameron, 57
Picton, 22, 28, 50, 59, 60, 71, 72
Platt, John Milton, 22, 32

Platt, Mary Elizabeth (wife of J. D. Silcox), 44
Platt, Willett McConnell, 44
Platt Street, Brighton, 44
Plomer, James, 60
Point Anne, 29, 36, 39, 52, 53, 54
Pope, John, 20, 32, 33
Port Robertson, 34
portage, 3, 4, 56, 61
Portage Road, Carrying Place, 13, 28, 71
Porter, E. G., 53
Porter, Thomas A., 62
Potts, Elizabeth (mother of Thomas Potts Powers), 65
Powers, Frances (wife of John May), 65
Powers, Thomas Potts, 65
Powles, Colin, xiv
Powles Report, xiv
Prescott, 20, 50
Presqu'ile Bay, 1, 2, 4, 7, 8, 10, 13, 14, 21, 23, 24, 28, 30, 32, 33, 35, 36, 40, 49, 50, 51, 52, 54, 65, 66, 67
Presqu'ile Point, 40
Presqu'ile Point Lighthouse, 6, 13, 49
Preston, Stanley of (The Governor General), 45
Prince Edward County, 4, 13, 22, 28, 32, 71, 72, 73
Prince Edward Street, Brighton, 71
Privy Council Office, 44, 45, 52
Proctor House Museum, Brighton, 26

Q

Queenston Heights, Niagara District, 12
Quinte Navigation Company, 50
Quinte West, City of, 70, 73

R

Railway and Canals, Department of, 18, 19, 20, 21, 22, 24, 32, 33, 42, 44, 45, 46, 47, 51, 52, 53, 57, 66, 68, 69
Rednersville, Prince Edward Co., 37
Richelieu and Ontario (Shipping Line), 50
Rideau Canal, 5, 6, 17
rip-rap, 38
Ritchie, Thomas, 43
Robertson, Scotty, 55
Row, John, 62
Rowan, J. H., 13, 14, 19, 32, 40
Rowe, Benjamin, 62, 63
Rowe, Mary Jane, 62
Royal Commission on Inland Navigation, 15, 17
Royal Mail (Shipping Line), 50
Royal Military College, 29
Rubidge, Frederick Preston, 20
Rubidge, Thomas Stafford, 20, 21, 23, 24, 26, 32, 51
Ruttan, Henry, 4

S

St. Lawrence River, 7, 14
Salt Point, 23, 49, 54,
Sanford, Dr. Charles M., 44
Saskatchewan, 25
Scriver, John, 67
Silcox and Mowry, 26, 28
Silcox, George, 44
Silcox, John David (J.D.), 23, 24, 26, 28, 29, 31, 36, 38, 42, 43, 44
Silcox, Joseph, 44
silver spade, 25, 26
Simcoe, John Graves, 1, 2, 8
Simcoe, Elizabeth, 1
Simcoe's Canal, 1, 4, 70
Smithfield, 56, 64, 68
Smithfield Road, 24, 56, 63, 65, 68
Smithfield Road bridge, 24, 28, 36, 39, 54, 56, 65, 69
Stage Coach King (William Weller), 29
Stanton, Sam, 67
steam dredges, 28
steam engine, 28
steam hammer, 34
steam whistles, 33
steamship, 9, 35, 41, 49, 50, 55, 67
steamship *Brockville*, 50, 55
steamship *Varuna*, 50, 55
Steuben County, New York, 12
Stoneburgh, Abraham, 64
Stoneburgh, Arnold, 68
Stoneburgh, Cadwell Ketchum, 64

Stoneburgh, Catherine (wife of Joseph Thomas Pelkey), 65
Stoneburgh, Clarissa (wife of William Stoneburgh), 65
Stoneburgh, Eliza Jane (wife of Joseph), 63
Stoneburgh, Joseph, 63
Stoneburgh, Peter, 63
Stoneburgh, Rachel (wife of Joseph Gould), 64
Stoneburgh, William (son of Peter), 63, 65
Stoneburgh's Cove, 13
Stoney Point, 61, 65, 66
Superintendent, 29, 42, 43, 47, 52, 67, 68, 69
Surprise, 43
survey, 2, 3, 4, 5, 6, 7, 13, 19, 20, 21, 26, 32, 39, 40, 53, 54, 55, 61
surveyors, 25, 26
swing bridge, 24, 30, 40, 47, 48, 49, 59, 60, 72, 73, 74, 75
Syracuse, New York, 23, 44

T

Talmage, Aaron W., 67
Taylor, George, 42, 43
Terry, Beatrice Ellen (wife of John Robert Harvey), 68
Thomson, Charles Poulett, 6
Thousand Islands, 50
trans-continental railway, 16, 19, 49
Trent Canal, 17, 34, 55, 67
Trent Port (Trenton), 6, 56, 70
Trent River, 6, 9
Trent-Severn Waterway, 73
Trent-Severn Waterway National Historic Site, 75
Trenton, 6, 8, 9, 25, 28, 31, 33, 34, 38, 43, 47, 50, 52, 53, 70, 71, 75
Trenton Bridge and Engine Company, 33, 34, 53, 54, 55
Trentonian, 38
Trenton Road, 12, 24, 28, 29, 51, 57, 62
Trenton Road bridge, 28, 29, 36, 38, 49, 55, 57, 58
Tupper, Charles, 18, 19, 45
Twelve O'clock Point, 6, 13, 21, 24, 35, 51, 52, 53, 70

U

Union Army, 11
United Empire Loyalist (U.E.L.), 63

V

Vermont, 15
Victoria, British Columbia, 16
Victoria College, Cobourg, 12

W

War of 1812, 3, 4, 63
War of Independence, 15
Webb, Adam Clarke, 25
Webb, Thomas, 25
Weddell, Robert, Jr., 31, 33, 38, 49, 53
Weddell, Robert, Sr., 33, 34, 36, 39
Weese's Creek, Presqu'ile Bay, 66
Welland, 23, 54, 57
Welland Canal, 5, 10, 15, 17, 23, 29, 44, 50, 60
Weller, Asa, 3, 4, 56, 61
Weller, John Laing, 29
Weller, William, 29
Wellers Bay, 1, 4, 6, 7, 10, 13, 14, 21, 22, 28, 61, 62
White, John, 35
Williams, Emily Jane (wife of Cadwell K. Stoneburgh), 64
Williams, Hester A. (wife of Charles Lee), 64
Wilson, John (father of Joseph), 64
Wilson, Joseph, 64
World War I, 50, 54, 55, 67
World War II, 69

Y

York (Toronto), 2, 56
York Road, 4, 56